94

The
SETI
Factor

How the Search for
Extraterrestrial Intelligence
Is Changing our View
of the Universe and Ourselves

The SETI Factor

How the Search for
Extraterrestrial Intelligence
Is Changing our View
of the Universe and Ourselves

Frank White

WALKER AND COMPANY
New York

Copyright © 1990 by Frank White

First published in the United States of America in 1990
by Walker Publishing Company, Inc.

Published simultaneously in Canada by Thomas Allen & Son
Canada, Limited, Markham, Ontario

Library of Congress Cataloging-in-Publication Data

White, Frank
The SETI factor : how the search for extraterrestrial intelligence
is changing our view of the universe and ourselves / Frank White.
Includes bibliographical references.
ISBN 0-8027-1105-7
1. Life on other planets. I. Title.
QB54.W48   1990   89-77858
574.999—dc20   CIP

Printed in the United States of America

2   4   6   8   10   9   7   5   3   1

To My Parents and My Children

Contents

Acknowledgments

Writing a book is a humbling experience. I always start the process determined to write the definitive work on a subject, and end up with the feeling I've only just begun to understand the topic.

I also find that the number of people who must be acknowledged for their contribution is truly staggering; my dependence upon others for the success of any undertaking is amazing.

The SETI Factor is not definitive, but it is a good start in grasping the immensity of this topic, and I can't thank those who assisted me enough, but I will give it a try.

Thanks must begin with my mother, Mary Anne, who gave me my first book on astronomy, and always encouraged me to realize my potential as a thinker. My dad, also named Frank, did research for the book, read and commented on the first draft, and remains a constant source of loving support for my work. My stepmother, Celeste, maintains an ongoing interest in my writing, and reads most of it as well.

My son, Josh, also does research for me, and continues to be a wise advisor on many aspects of my writing. My daughter, Ruth, is unwavering in her interest and support for my work, and in disseminating it to the world.

Beyond my family, others who deserve special recognition include Charles Redmond, Will Thorndike, and Larry Marschall. It was from Charles Redmond, a public affairs officer with NASA, that I first heard about the NASA SETI project, and realized how imminent contact might be. As my editor, Will Thorndike's enthusiasm for the project, thoughtful editing, and intellectual contributions to the concept have been invaluable. Larry Marschall, professor at Gettysburg College, diligently read the first and second drafts of the manuscript and made extensive comments that measurably improved the work.

Others who read and commented on various drafts of the manuscript included: Bob Arnold, Michael Benes, Bruce Campbell, Dennis Campbell, Andrew Chaikin, Gary Coulter, Richard Tresch Fienberg, David Harper, Rena Shulsky, Bob Strayton, and Frank C. White.

Then there are all those who were interviewed for the book. Space simply prevented my using the comments of everyone, but all the interviews were extremely valuable, and my thanks to these interviewees: Poul Anderson, Bob Arnold, Isaac Asimov, Peter Backus, David Baron, Greg Barr, Kelly Beatty, John Billingham, Ted Bonnitt, Bruce Campbell, Andrew Chaikin, Pam Colorado, Rick Cook, Kent Cullers, Leonard David, Hugh Downs, George Field, Ben Finney, Louis Friedman, Jim Funaro, Jim Gorman, Willis Harman, David Harper, Lynn Harper, Richard Hoover, Paul Horowitz, Eric Jones, Mary Jones, Larry Kaye, Michael Klein, David Latham, Leroy Little Bear, John Logsdon, Michael Michaud, Philip Morrison, Bernard Oliver, Michael Papagiannis, Jesco von Puttkamer, Charles Redmond, Richard Robbins, Tony Rothman, Carl Sagan, Bruce Shackleton, Don Tarter, Jill Tarter, Charlie Walker, John Wilford, David Williamson, and Tom Zane.

The final production of a book is a real challenge for everyone involved, and I want to thank Tom Carey for his thoughtful job of copy editing, and Helen Driller for taking the manuscript to final completion.

I must also thank Bob Carr once again for supplying me with the Wangwriter that has now produced its third book.

Thanks also to everyone working at the NASA SETI office, the SETI Institute, Project META, and Eleanor York at Cornell's Laboratory for Planetary Studies.

There are many people in my network of friends and family, far too numerous to name, who have consistently encouraged me in my writing. A few representative names would include: Allen, Barbara, John, Madeline, the two Nicks, Prilly, Randy, Susan, and all my friends at Hill and Knowlton. Thanks to all of you and everyone who cared about this book.

It goes without saying that while I do want to thank everyone for their help on The SETI Factor, I am solely responsible for its content.

Credits

The author is grateful for permission to quote from the following works:

Preface

The Overview Effect, my first book about space exploration and human evolution, analyzes the short-term and long-term impact of seeing the earth from the vantage point of outer space. The book concludes that the primary impact of space exploration is on human consciousness, and is a key to human and universal evolution..

Writing *The Overview Effect* led to the realization that other aspects of space exploration would cause shifts in our understanding of the universe, and that contact with extraterrestrial intelligence would surely be a moment of tremendous importance in our history. Making contact with a galactic civilization (Galaxia) appeared to be one of a series of changes in awareness being brought about by space exploration.

The question is, how can humanity take advantage of the opportunities offered by space, consciously using it to shape our evolutionary direction?

In a concluding chapter of the book, I posed an answer in the form of the "Human Space Program," a global commitment to exploration of the universe lasting through the next millennium. In addition to defining a vision, purpose, goals and objectives for the program, twenty start-up projects were listed.

Two of these projects related directly to the search for extraterrestrial intelligence:

15: *Investigate the probable impact of contact with a galactic civilization:* Contact without preparation could be devastating, but thoughtful contact can be of great benefit to both species. Active efforts to prepare for communication may hasten the day when it occurs.

16: *Begin a campaign to inform humanity that we are in space:* Increase awareness in general, especially of the preparations for contact with Galaxia.

After completing *The Overview Effect*, I conceived of a series of books, each of which would lay the foundation for one or more of the twenty projects described as the initial steps in the Human Space Program. I discussed the series with Will Thorndike, a Senior Editor at Walker and Company, and he expressed enthusiasm for the idea; Will especially liked the concept of a book about the scientific search for extraterrestrial intelligence.

After a series of meetings, we crafted the basic outline for what eventually became **The SETI Factor.** Like all books, it has gone through many transformations since those early days. Still, it remains true to the fundamental idea of informing people about an issue of immense importance so that they can take appropriate action on it.

Much of *The SETI Factor* is about the potential impact of contact with extraterrestrial intelligence. Until that day, I hope that the book will fulfill its purpose and have a significant impact on terrestrial intelligence!

—Frank White

Chapter One

What's at Stake?

> *Our obligation to survive is owed not just to ourselves but
> also to that Cosmos, ancient and vast, from which we
> spring.*
>
> —Carl Sagan, *Cosmos*

WHAT IS AT STAKE IN THE SEARCH FOR EXTRATERRESTRIAL Intelligence (SETI) is nothing less than our understanding of what it means to be human. SETI challenges us to come to terms with our identity and purpose in this vast universe of which we are a small but important part.

Human beings learn and grow by asking questions about the nature of things and then setting out to find the answers. The questioning begins with the small child asking, "Why?"; and, for the truly curious adult, the process never ends. The questions are at the heart of science and exploration, as well as personal growth and social evolution. The bigger the question, the harder it is to find the answer, and the greater the payoff. It's like a diving competition— competitors get points for how well they perform the dives, but there is also a degree-of-difficulty rating. The harder the dive, the more points are awarded.

We form our identities by comparing ourselves to other people and societies. Without an idea of others, there can be no image of ourselves. "I" am everything that isn't the "other," which can be another person, society, the universe, extraterrestrials, or God.

The more comprehensive our idea of others, the more we can learn about ourselves, and the greater the expansion of our identity. For example, if I ask myself, "What is my role as a citizen of my country?" that is a significant question and it makes me think deeply about myself. If I ask, "What is my role as a citizen of planet Earth?" the stakes are higher, and the questioning more profound. But if I ask, "What is my role as a citizen of the universe?" the question has become as broad as it can be, and I'll have to think long and hard to come up with an answer.

The search for extraterrestrial intelligence may hold the key to becoming citizens of the universe because it will yield so much knowledge about the nature of the universe itself. In asking how the universe is put together, we are trying to take what is now mysterious and unknown and turn it into useful knowledge. We can't be good citizens of our country without understanding how our country works, what are its values and traditions. In the same way, we cannot take on this larger citizenship without a vast amount of new knowledge.

At the same time, since we aren't really separate from the universe, everything we learn about what is "out there" tells us new things about what is inside ourselves. The inner and outer search are not as separate as they sometimes seem.

For all these reasons, one of the major questions facing human beings today is, "What is the nature of the universe and what is the role of human beings within it?" It's stated as a scientific question, but touches on theology, philosophy, psychology, and most other disciplines. It is ultimately a question of cosmology, a term that means "knowledge of the cosmos."

To living, thinking beings, the question focuses on whether life and intelligence are common or rare in the universe. What we are really asking is, "Are *we* common or unique?" It's the same question we ask ourselves as individuals, and it is certainly one of the most important in our lives.

The effort to answer this question in unambiguous scientific terms makes SETI exciting and vital. Answering the question, "Are we alone in the universe?" is a major project, and the answer is elusive, as shown by the fact that humans have been working on it for several thousand years now. However, the payoff is clearly enormous, a fact that has become increasingly clear to many experts in the past several years.

For example, astronomer/author Gerald S. Hawkins believes that

SETI may lead to one of those "mindsteps" that are milestones in human evolution, the most recent having been triggered by the launching of the first rockets into outer space.[1]

Author Steven J. Dick suggests that SETI will either confirm or contradict the cosmology that has been dominant since around 1750. This "bioastronomers' cosmology" assumes the laws of evolution and nature to be the same throughout the universe, the building blocks of life to be plentiful everywhere, and a universe teeming with intelligent life to be highly likely.[2]

Michael A. G. Michaud, Director of the U.S. State Department's Office of Advanced Technology, is a leader in the effort to prepare for the social consequences of contact. He believes that contact and cooperation among intelligent life forms may be the key to survival when the universe begins to die as the energy of the stars is expended and the cosmos becomes a cold place populated by white dwarfs and black holes.[3]

Ben Finney, Professor of Anthropology at the University of Hawaii, suggests that SETI is far more than an intellectual exercise. He says that SETI, and all other forms of space exploration, are part of an unconscious but ongoing "reconnaissance mission" for the human species. As an expansionary "exploring animal," humans want to know whether there is anyone out there because of the practical effect of that knowledge on our long-term future.[4]

SETI, then, is far more than an intellectual curiosity for a few astronomers—it may be one of the most important tasks that human beings have ever undertaken. As we approach the end of the twentieth century and the beginning of the Third Millennium, there is a growing sense of expectation among some members of the scientific community that, if current trends continue, we may within our lifetimes find an answer to the question that has puzzled the best minds on our planet for so long: "Are we alone?"

Note that it is the *scientific* search for extraterrestrial intelligence that is now gaining momentum. The "SETI Factor" has been a force in human development for thousands of years, and it represents a deep human need to feel connected with the cosmos, rather than isolated and alone. However, this concern does not always manifest itself in scientific terms. It can be expressed as religious feeling, through art or music, and as philosophy and mythology.

In the past, extraterrestrial intelligence might have been seen as angels, gods, or God. In modern times, it may be seen as aliens piloting UFOs. Recently, people have tried to merge these two streams of thought by suggesting that the gods of mythology were extraterrestrial astronauts, or that many of the descriptions from the Bible represent previous visits to Earth by aliens.

The scientific community has remained skeptical of these claims. Science requires proof, and its criteria have not yet been met by most claims for the existence of extraterrestrial intelligence. A scientific confirmation would add a new voice to the dialogue, transforming the discussion and taking it into a new realm for society as a whole. For the first time, a substantial community of skeptics would move unreservedly into the camp of the convinced.

The current phase of the scientific search for extraterrestrial intelligence is just over thirty years old, having begun with a seminal paper written by two physicists, Philip Morrison and Giuseppe Cocconi, in 1959. Along the way, there have been many benchmarks and watersheds, such as the first search using radio telescopes, undertaken by Dr. Frank Drake in 1960; a series of search efforts in the 1960s and 1970s; and the inauguration of Project META (Megachannel Extra terrestrial Assay), the largest privately-funded SETI project, in 1983.

The decade of the 1990s promises to bring SETI to a new level of funding, opportunity, and public interest. On Columbus Day, 1992, the National Aeronautics and Space Administration (NASA) plans to initiate the largest and most sophisticated search ever carried out, the Microwave Observing Project (MOP). Using radiotelescopes and newly developed computer technology, the project will begin scanning the sky for signals sent to us by civilizations beyond the bounds of our solar system.

Within six years, this project will have conducted a "targeted" survey of some eight hundred stars to one thousand within eighty to one hundred light-years of Earth, and a "sky survey" of the entire Milky Way galaxy. According to NASA, the project will have generated more information in its first half-hour of operation than all other searches during the past three decades combined.

The effort is ambitious, but it is only one of many attempts to determine whether the blossoming of life and intelligence on Earth is a unique event, or a common occurrence. During the same period, Project META will be expanded to conduct observations from the Southern Hemisphere, and its current systems will be significantly upgraded.

The implications of this work are enormous, regardless of the result: The answer could be, "No, we are not alone. There are other beings of comparable or greater development in our galaxy and in other galaxies." We may then have to learn how to function as part of a galactic or intergalactic civilization. Using Ben Finney's reconnaissance metaphor, our scouts may return to base, and tell us, "Lots of campfires out there." If so, human beings may have to fit themselves into an existing structure, becoming students of mentor-civilizations eons older than our own. The universe will become our "university," as we find ourselves in a "galactic kindergarten."

The answer may be, "Yes, we are alone. There are many wonderful things out there, but nothing quite like us has appeared in our galaxy or elsewhere." If that is the case, then we would be facing an empty universe, devoid of life and intelligence as we know it. Then, too, life and intelligence may abound, but we are by far the most advanced forms; we are the mentors.

Under those circumstances, we will be confronted with the burden of preserving a precious spark of awareness, disseminating it throughout the cosmos, and nurturing other sparks into the light of knowledge. What will we do with this opportunity; how will we handle this awesome responsibility?

The search for extraterrestrial intelligence, regardless of its outcome, is an identity-expanding process for humanity. Thinking about SETI means thinking about our planet, the universe and ourselves.

SETI touches on most fields of human knowledge, raising so many fascinating questions that no single book can cover them all. In this book, we will focus on a few key concepts:

1. **The search for extraterrestrial intelligence, described in different ways, has been with us since the dawn of human consciousness**
2. **This "SETI Factor" has played a key role in human evolution for millennia**

3. The social and psychological impact of the search for extra-terrestrial intelligence will be substantial, regardless of the findings
4. The time has come to "get ready for SETI," by exploring how we think about our relationship with the universe as a whole

As the Third Millennium approaches, human beings are leaving Earth both physically and mentally. The SETI Factor is one more manifestation of our reaching out to a cosmos much greater than the single planet of our birth. Regardless of the answer to the question, "Are we alone?" the challenge is to become true "Citizens of the Universe," a species actively engaged in a domain far larger and more complex than our home planet, a species with a new and more comprehensive cosmology.

This book is about that challenge and how we can respond to it.

Chapter Two

Framing the Question

> SETI is based on the compounding of the probability of several events occurring: you have to get the right star and the right planet, the planet has to have life on it, the life has to evolve intelligence, they have to turn their radios on, and then they have to live long enough for us to find them.
>
> —Dr. Kent Cullers,
> NASA SETI Project

BACK IN THE 1950S, A POPULAR SONG SAID, "I'VE BEEN SEARCHIN' every which a way, yea, yea . . ." And they knew what they were looking for: "I'm gonna find her," and "You know I'll bring her in."

SETI isn't like other searches, whether it is for the perfect relationship, a quarterback who can throw for long yardage, or a house in the suburbs for less than $100,000. We usually search for things that we know are there, or at least think that there's a high probability that they exist.

SETI is a search for something that might *not* be there; success will simultaneously prove that the object of the search exists and the method of looking for it was correct! Extraterrestrial intelligence is the object of the search, but what is it exactly? If we analyze SETI carefully, it forces us to ask, "What are we looking for?"

Answers are shaped by the way questions are framed. Scientists know that forming the question affects the experiments they set up to validate hypotheses. Sometimes, experiments work, but don't answer the question. Sometimes, they go wrong, but produce new discoveries never before imagined.

Most of today's high-visibility SETI activities focus on trying to detect signals from civilizations in other star systems. The existence of these civilizations is assumed, based on one piece of evidence: ourselves; otherwise, they are purely hypothetical. While this may be the best methodology available today, it is not all-inclusive.

As Dr. Kent Cullers, who is in charge of signal acquisition for the upcoming NASA SETI project, has said:

> Why are we starting at the top, looking for intelligent life? The reason is because that search could pay off tomorrow. It is playing a long-shot scientific intuition, but a good scientist only plays his or her intuition so far. If we keep expanding the radio search and get to some point where the cost of the radio search is so high that you could actually do other things, such as building big telescopes to directly image other planets, or rockets to go to those planets, then you should consider those modes also.[1]

It will be very convenient if extraterrestrials use the electromagnetic spectrum as we do, because those are the kinds of signals we can detect. However, civilizations advanced enough to communicate across interstellar distances may now be using techniques far beyond what we know. Under those circumstances, they might be talking up a storm right under our noses and we would never know it.[2]

Imagine an isolated tribe on Earth *circa* 1990 listening for drum signals as an indication that there is intelligent life beyond their territory. Hearing nothing they could recognize, they might agree among themselves, "Nobody's out there," while invisible radio and television signals zip through the air all around them.

On the other hand, as author Isaac Asimov points out, the current approach does have a logic to it. Interstellar transportation appears to be time-consuming and expensive, and most forms of radiation are absorbed by the interstellar medium. Therefore:

> We're attempting to study the universe in detail, looking for microwaves that are not random in nature nor regular. Neither randomness nor regularity carry information. We need something that is irregular, and not random. That would indicate the presence of intelligence.[3]

The scientific community is not wholly united on the appropriate strategy for the search, though most agree that it is reasonable to undertake the radio search in the short term. However, we should also maintain a very broad long-term perspective, to avoid missing major opportunities in the future. To ensure that the broader perspective is there, let's take a look at the term "search for extraterrestrial intelligence," and think about the assumptions behind that phrase as it is used today.

The Search

The term "search" is defined in the dictionary as making "a thorough examination in order to find something; explore."[4] The radio search does involve looking for something, and it is a form of exploration. However, a broad definition of the term can include many activities not always linked with the search for signals.

For example, space exploration using unmanned probes or manned spacecraft is usually justified for reasons other than a search for life and intelligence. However, the SETI rationale is often present in the foreground or background of a mission. The search for life was central to the 1976 Viking mission to Mars, which included three experiments explicitly designed to test the Martian soil for living organisms. The experiments produced ambiguous results, and the status of life on Mars remains uncertain. Most scientists now believe the planet to be lifeless, but the *Viking* mission also suggests that Mars once had an abundance of water, which is essential to life as we know it.[5]

Almost no one expected to find intelligent beings on Mars, but searching for life and even signs of intelligence has become one argument in favor of an international manned mission there in the near future. The discovery of such remains anywhere in the solar system would transform our views of the existence of extraterrestrial intelligence elsewhere, even though we would not be in direct communication with that intelligence.

SETI could logically include such searches being undertaken on Earth. The problem is that initial claims regarding "ancient astronauts" have failed to follow accepted standards of scientific inquiry, and have therefore been discounted by the scientific community. In addition, there is always uncertainty regarding any terrestrial discovery concerning how it can be shown to have certifiable extraterrestrial origins. By contrast, anything found on Mars, or any other planet would almost certainly be "extraterrestrial."[6]

Similarly, interest in the *Apollo* missions to the moon focused on the incredible achievement of sending intelligent terrestrial life there and bringing it back safely. Because of the nature of the political atmosphere at that time, American planners and politicians also emphasized the goal of "beating the Russians" in a "space race." However, the scientific elements of the mission included collection of moon rocks to learn more about the origins of the moon, and to determine if the moon were biologically active. While most scientists doubted that it was, the *Apollo* astronauts from the first three landings were quarantined on their return, in case they had picked up any extraterrestrial viruses or bacteria.[7]

Nearly every report or analysis of exploration within the solar system includes a statement about whether the mission will add to our knowledge of the prevalence of life and/or intelligence in the universe. The unmanned *Voyager* probe, in particular, created quite a stir with some of its findings among the outer planets.

Voyager confirmed that Titan, a moon of Saturn, possesses an atmosphere not unlike that of the early Earth, with many organic compounds present.[8] Europa, a moon of Jupiter, appears to be covered with an ocean of ice, and it is tantalizing to wonder what might be under it. Io, another moon of Jupiter, is extremely active geologically, and we now have pictures of volcanoes dramatically erupting on its surface.[9] Neptune's moon, Triton, also supports an atmosphere, and other fascinating features that have yet to be explained.[10]

The solar system as a whole looks far more complex and full of surprises than we expected. While most planetary scientists still believe that Earth is the only planet harboring life and intelligence, the variety of circumstances in the solar neighborhood makes all assumptions far less certain than before.

Human beings have been exploring outer space physically for only a few decades. During all that time, those involved in the exploration might give different reasons for their interest, such as increasing human knowledge, founding new societies, creating wealth, and enhancing national prestige. However, it would not be going too far to say that not only is SETI a form of space exploration, but space exploration is also a form of SETI. Even space settlement, which is concerned with expanding the human presence into the universe, may be the unintended agency for the discovery of extraterrestrial life and intelligence—part of a grand "reconnaissance" for humanity being undertaken in many different forms.

Perhaps the most important aspect of the word "search" is the element of uncertainty it introduces. When we are searching for something, we admit from the outset that we don't know *where* it is. But do we know *what* it is? The definition of "search" points to the question of what that "something" is.

Extraterrestrials

At first, the term "extraterrestrial" seems obvious. It must mean a being with origins and existence outside the boundaries of the earth. It isn't that simple, however.

When popular culture speculates on what an extraterrestrial might be like, it produces a creature like "E.T." E.T. is not so far from the kind of life form that a scientist might imagine evolving on a warm planet (no need for fur, feathers, or clothes) with somewhat greater gravity than that of Earth (stocky body slung low to the ground), a nitrogen/oxygen atmosphere (no requirement for special breathing apparatus when exploring terrestrial environments) and an internal digestive system similar to our own (ability to consume beer, Reese's Pieces, etc.).

At first, E.T. seems radically different from a human being, but on reflection, the similarities are more striking. E.T. has two eyes, a nose, a mouth, two arms and two legs. He is not human, but he is humanoid. He is close enough to being human that he even learns to speak English! He does have special powers—he can fly and heal people with the touch of his finger. That's outside the range of what normal people can do, but there are stories of people who can do some of those things. Overall, E.T. is strange but well within our comprehension.

On the other hand, a definition of "extraterrestrial" that includes *any* life form originating off the planet Earth opens up a stunning range of possibilities. These possibilities expand the definition of the term "extraterrestrial," and lead to many new questions.

When we think about extraterrestrials that are not built on the human model at all, everything gets complicated. Are there planetary conditions that might support a creature without a head, two arms, two legs, two eyes, a nose and a mouth? Could something be alive and yet not be organic, like a machine? Must extraterrestrials evolve on a planet, and must they be in a physical body? Allowing for millions of years of evolution beyond our own stage of development, and for life in space, who knows the evolutionary direction extraterrestrials might take?

Charles Walker, Special Assistant to the President of McDonnell Douglas, and President of the National Space Society, has flown on the space shuttle three times as a Payload Specialist. He says his experience suggested that the nature of a spacefaring extraterrestrial might be hard to imagine:

> Space is a rigorous place in which to travel, live, and work, and it is going to take more than a casual interest on the part of a race to develop technologies and the continuity of will and spirit to move into space initially. The environment there is a different one than we have here on the surface of this planet, and it will certainly drive evolution rapidly, because it is so different. In order to accommodate it, we will have to change or change ourselves in a fairly rapid manner.[11]

When human astronauts go into space, their bodies react to the space environment, responding dramatically to weightlessness. In a few days' time, the body gets longer, legs grow thinner, faces fill out, the heart beats more slowly, calcium is shed from the bones, and muscles begin to atrophy. What will happen over an extended period, or to children born in space? The adaptations may well continue at an accelerated rate with each new generation.

Conscious adaptation, using genetic engineering or other techniques, is another story altogether. The creation of "cyborgs," or human/machine hybrids, might be appealing in space, which is inhospitable to organic life. When we spend extended periods of time on other planets, such as Mars, we will either have to change those planets to accommodate our biochemistry, or ourselves to adapt to the planet's conditions. If that is so for our species, why wouldn't it be the same for others?

Ultimately, as a species moves off its home planet, it may "speciate," or produce a totally new species in response to the outer space environment.[12] If that were to happen with extraterrestrials, the result could be an organism that looks and thinks very differently from humans because it initially evolved in an alien environment (its home planet) and then adapted to an even more alien environment (outer space).

It isn't possible to predict the kinds of organisms that would be produced after millions of years of evolution in outer space. Our own descendants will certainly be dramatically different from us, as humans become a multi-planet species. Such thoughts suggest that calling a spacefaring species from another planet "alien" is an appropriate use of the term. However, after terrestrials and extraterrestrials have both been in space for thousands of years, we may look more alike than different!

Intelligence

Intelligence is a quality that is hard to grasp even without bringing extraterrestrials into the picture. The dictionary defines it as "the capacity to acquire and apply knowledge; the faculty of thought and reason."[13] The definition is simple, but the question is, "Who has that capacity?"

For many years, it was assumed that other animals lacked intelligence, operating purely according to instinct. Researchers are now challenging that view, and many are willing to ascribe intelligence not only to higher mammals such as dolphins and chimpanzees, but also to dogs and cats. However, if intelligence is granted to them, where will the process stop? How will the human understanding of this faculty change as work continues in the field of "artificial intelligence," which promises to produce machines that mimic the human thought process?

The "search for artificial intelligence" and the "search for extraterrestrial intelligence" have much in common in that they are both intimately bound up with issues of human identity. It is not surprising that AI (Artificial Intelligence) and SETI originated at about the same time (the late 1950s), that many of the same people are active in both fields, and that science fiction writers have been fascinated almost equally by extraterrestrials and robots. For anyone concerned with the question, "What does it mean to be human?" both fields offer room for tremendous speculation.

The relationship between AI and SETI is more direct, however. As researchers provide computers with greater capability, "cyberphobes" try to make the definition of intelligence increasingly restrictive, maintaining the human position of superiority over computers. "Cyberphiles" at the other extreme tout the computer as the next great step in evolution, a new kind of species that will eventually surpass human beings. Experts on computers and robots talk about "downloading," shifting the entire memory and consciousness of a human being into a computer; what is non-physical about the person would live on after the physical body dies.[14]

From our perch on the evolutionary ladder today, ideas like "downloading" may seem absurd, and even dangerous. However, any idea of how intelligence might evolve in the future is likely to be overly limiting. If evolution is correct, the human beings of today are descended from the fish of yesteryear. Human beings evolved as an adaptation to the land environment, and are quite well suited to it. But if you could have stood on Earth eons ago, and looked at fish in the primeval oceans, how could you have predicted that humans would be the result of the evolutionary process you were observing?

Human beings are coping with life in a highly toxic Earthly environment, and are contemplating life in outer space. Natural evolutionary pressures cause unpredictable adaptations in the normal course of events. Today, genetic engineering and artificial intelligence offer the opportunity for conscious adaptation, both physically and mentally.

The classical evolutionary view assumes that intelligence first appeared in human beings who had evolved into a big-brained species separate from their ape ancestors. On the other hand, intelligence itself may be a different phenomenon from what we imagine. Perhaps it is better to define it in another way altogether. There are those who would say that ants and bees have a kind of group intelligence, that micro-organisms sometimes appear to be intelligent, that even planets might have a form of consciousness.[15]

Might a form of intelligence evolve that is not contained in a physical body; could it be like a swarm of micro-organisms floating through the interstellar regions? The astronomer Fred Hoyle explored this option in a science fiction piece about something called "The Black Cloud."[16]

Extraterrestrials *may* exist in almost any form, and might behave in truly unpredictable ways, but our experiments are not designed to detect every possibility. The search is actually not "SETI," but "SETILO"—a "search for extraterrestrial intelligence *like ourselves.*"[17]

In the "Assumption of Mediocrity," enunciated by Dr. Carl Sagan and others, it is assumed that all of the universe is constructed similarly, there are no "privileged sites," and that evolution follows similar paths in local regions. This leads us to the idea of an extraterrestrial civilization that develops technologies like our own. Having developed such technologies, they then use them to contact other societies like themselves. It's something humans do, and the behavior is projected onto extraterrestrials.

Such a scenario makes the SETI enterprise believable, but it may exclude the possibility of contacting an extraterrestrial intelligence that has developed in some truly exotic fashion not readily conceivable to human beings.

Such intelligence might evolve to such a level that we would be unaware of it, and it might be unconcerned with us. Terrestrial examples prove that evolution creates enormous gaps between living creatures on the same planet. As I walk around my apartment each day, I pass in front of our aquarium and look at the swordtails and black mollies swimming inside. As far as I can tell, they have little or no awareness of me, and I doubt that they have a conceptual understanding of what life is like where I am. And is my comprehension of life at their level any better than theirs of mine? Can we communicate?

We are contemplating contact with "extraterrestrial intelligence" without an explicit definition of "extraterrestrial intelligence." Add to that uncertain foundation the fact that intelligence might evolve in radically different ways on other planets, taking on totally unexpected forms in outer space, and the need for definition is clear.

However, the definitional discussion begins to seem endless, and the question becomes, "Is all this effort really necessary? What's the point, other than a philosophical exercise?" It *is* important in understanding the limits of the "truths" our experiment is likely to produce. It is quite possible for human beings to launch a search for extraterrestrial intelligence, carry out an extensive project, and then announce that nothing has been found: therefore, there are no extraterrestrials. This conclusion would have dramatic consequences for the future of human evolution, but it might also be wrong.

The universe is enormous, and the announcement that nothing had been found might mean that only a small portion of it had been searched. But suppose nothing had been found because nothing had fit into a predetermined, unspoken definition of extraterrestrial intelligence, i.e., a civilization enough like our own to be sending out signals that we can understand? It is possible, even if unlikely, that extraterrestrials are all around us at this very moment, but because they are not behaving according to our assumptions, we cannot "see" them.

Today's SETI projects are highly unlikely to detect any exotic forms of intelligence. They are structured so that they will find "ETLO" (Extraterrestrials Like Ourselves), or creatures at least somewhat like humans. Yet, the history of science is a story of major breakthroughs occurring by accident. A SETI experiment may fail to achieve its stated goals, yet stumble onto something far more profound. If we ourselves are truly intelligent, we will be open to this possibility.

Detection Defined

Basically, detection of extraterrestrial life and intelligence requires discovering processes at work in the universe that cannot be explained as natural phenomena, or as phenomena caused by terrestrials. Detection legitimately encompasses investigating meteorites that have crashed into the earth to determine if they carry organic compounds, and it can include any evidence of past or present human contact with extraterrestrials. The only criterion is that the scientific method be utilized, i.e., evidence cannot be skewed to fit a pre-existing point of view, but must be measured in terms of its validation of a hypothesis.

Within our solar system, detection can include all probes or manned missions sent to other planets to look for past or present life and intelligence, using the same criteria. Within our galaxy and throughout the universe, less direct methods are required, but the same basic approach remains valid. The question is always, "Given what we know about the natural order of the universe, where do we find phenomena that don't fit?"

As astronomers and astrophysicists develop models of the universe, and compare those models with observations, they frequently observe contradictions that are not easily explained. As a rule, they do not invoke intelligent life to explain these contradictions because we are so uncertain about its existence. However, it would be fascinating to go in the other direction, and use the assumed existence of such intelligence as an explanation.[18]

Dr. Otto Struve, a distinguished European astronomer, wrote and spoke on this issue as early as the 1950s. Saying, ". . . we ourselves are now capable of producing at will various phenomena . . . which could be observed from distant planets . . . . We must, therefore, revise our thinking and incorporate in our theories possible effects of the free will of other living beings . . ."[19]

Expanding Our Thinking

SETI is most meaningful in the context of the Western scientific paradigm, which is not the only available model of reality. When other models of reality are used, the search itself may become meaningless.

For example, the thinking of Native Americans is holistic, ascribing life and consciousness to everything in the universe. The critical difference between native thinking and the Western model refers back to the issue of identity; native thinking does not make a strong distinction between the observed and the observer, between self and other.

Standing apart from the universe and performing experiments on it in order to confirm hypotheses is foreign to that way of thinking. The search for extraterrestrial intelligence would be unnecessary, because the object of the search is found wherever we look!

To the modern mind, such an attitude may seem naive and simplistic. However, the separation of observer and observed, and the resulting emphasis on separate, individual identity has been achieved at a heavy cost. It allows us to see the universe as an object to be acted upon, to manipulate and exploit for human purposes, but it also creates an alienation that is not healthy.

The "alienation factor" raises the question of whether SETI, at its deepest levels, may be an effort to achieve a new kind of connection with the universe, working within a framework that is acceptable to the Western scientific model. Perhaps SETI is an acceptable way for us to seek that re-integration, a feeling of connectedness which has been shattered by standing apart from the cosmos and examining it as something that is not alive, not intelligent, and separate from ourselves.

Clarifying the Assumptions

SETI is exciting because it raises fundamental questions. Assumptions may not always be clear, but they will evolve with the process. Moreover, if the assumptions behind the search for extraterrestrial intelligence are not settled at the moment, the assumptions behind this book can be made clear. Here, it will be explicitly assumed that the current effort is a search for intelligence enough like ourselves that we can comprehend it. While that may be a limiting approach, the payoff of success will still be enormous and it is an appropriate way to focus the effort.

The form of that intelligence is not the critical issue, but rather its ability to communicate with us. If it is a black cloud or a cyborg, that doesn't matter, as long as the cloud or cyborg can communicate with human beings and is willing to do so. However, the communication must be something that most humans can agree to be valid; it cannot be a "psychic transmission" to one person or a group of people.

The communication must also be open, public, and easily verifiable. The very fact that there is a "UFO debate" disqualifies UFOs according to this criterion. That doesn't mean that UFOs aren't real, or even that they aren't extraterrestrial spaceships, only that their alleged communication with humanity is not open, public and easily verifiable.

If a UFO lands on the White House lawn and extraterrestrials emerge and say, "Take us to your leader," then the UFO debate will be over and so will the search for extraterrestrial intelligence. Until then, the search for distant planets and radio signals emanating from them is the "only game in town" because success will be easily verified and agreed upon. While this may seem to be a narrowing of scope, it is an enormous undertaking and will trigger a profound impact on human society—that is the ultimate subject of this book.

No one can predict how the current SETI experiment will turn out; until contact occurs, we can know very little about the extraterrestrials. However, we do know, and can learn much more, about terrestrials and how they respond to the results of the experiment. From our own point of view, that is where our responsibility lies, and that is the area in which we can do the most for ourselves and for the universe.

Chapter Three

The Debate

As a species, we now have the power to govern evolution.

—Lynn Harper, former
NASA SETI Program Manager

L ISTENING TO THE DEBATE OVER WHETHER LIFE AND INTELLIGENCE
are abundant or scarce is like sitting in on a courtroom trial.
The lawyer for the prosecution sways the listeners in one direction
by the force of his argument, as evidence against the defendant is
piled up, point by point. At the end of the dissertation, it seems
impossible to take another point of view seriously. Then, the defense
attorney rises and presents a totally different perspective, bringing
out facts that the prosecutor had ignored or skimmed over, and the
prosecution's case doesn't look quite so strong. What seemed incon-
trovertible is tinged with doubt and uncertainty. The truth must be
lurking somewhere, but where, exactly, is it?

The difference between the SETI debate and a court case is the
difference between natural and human law. In court, the laws are
known to all parties because they are human inventions. The debate
is not over what the laws are, but how they apply to a specific
situation. In the SETI debate, the argument is about what the laws of
the universe really are: Does the universe produce life and intelligence abundantly, originating in many locales, or does it have a
different strategy, proceeding from one, or a very few, source points
and spreading out from there? The debate is valuable precisely
because an examination of the "case for" and the "case against"
provides an illuminating journey into human understanding of the
universe.

The Case For

The case for abundant life and intelligence was first stated thousands of years ago, based on assumptions that have changed very little over time. In fact, until recently, the speculations of scientists of the late twentieth century AD were not very different from those of the Greek philosophers of the fourth century BC, who were also fascinated by the SETI Factor.

In both cases, the assumption that there are many planets in the universe harboring a variety of intelligent life forms (known as the "plurality of worlds" argument) was based primarily on two facts:

Fact #1: Our sun is a star, and it is accompanied by planets, at least one of which (Earth) has produced an environment conducive to the evolution of intelligent life.

Fact #2: There are many stars in the universe.

It is not only the early philosophers and modern scientists who see the implications of these two facts. People who have given only cursory thought to the matter will say, "We can't be the only ones. It just doesn't make sense when you look at the numbers."

Observations over time appeared to confirm the case for abundance. Within our own solar system, more planets were found (and in recent years, more moons), and everywhere we looked, there seemed to be more stars. Then, in the early twentieth century, galaxies were recognized as systems of billions of stars separate from our own Milky Way galaxy. It is now known that *Fact #2* can be expanded to state that there are billions of stars in our galaxy and billions of galaxies in the universe.

"Common sense" seems to dictate that there must be planets around some of those other stars, and that these planets must have evolved life and intelligence.

Common sense isn't always right, however, and it isn't necessarily scientific. To meet the challenge of formalizing the abundance argument, the two basic facts became the foundation for the "Assumption of Mediocrity," which is generally associated with Dr. Carl Sagan.

The Assumption of Mediocrity is built on the unflattering notion that the Earth, sun and solar system are pretty ordinary. It further supposes that the building blocks of life are common throughout the universe, and the laws of evolution are the same everywhere. If all of this is correct, then the vast numbers of stars are meaningful—they offer many breeding grounds in which the process observed on Earth can be repeated.

This viewpoint is also called the Copernican Principle, which states that "There are no privileged sites in the universe." It was Copernicus's heliocentric model of the solar system that overthrew the Ptolemaic model, in which the Earth was privileged indeed, i.e., the center of the universe.[1]

The vast numbers of stars are important to the Assumption of Mediocrity for another reason. The Assumption does not require divine intervention or any special event to produce life and intelligence. Rather, the working of chance in an evolutionary framework is deemed sufficient to produce many sites in which life and intelligence might develop.

In the 1960s, Dr. Sagan and a Russian collaborator, I. S. Shklovskii, wrote a pioneering book entitled *Intelligent Life in the Universe*. In discussing the Assumption of Mediocrity, they wrote:

> The idea that we are not unique has proved to be one of the most fruitful of modern science. The atoms on Earth are the same in kind as those in a galaxy some 5 to 10 billion light years distant. The same interactions occur, the same laws of nature govern their motions. . . .[2]

Compare the ideas of Sagan and Shklovskii to those of the Greek philosopher Epicurus (341–270 BC):

> There are infinite worlds both like and unlike this world of ours. For the atoms being infinite in number . . . are borne far out into space. For those atoms . . . have not been used up either on one world or on a limited number of worlds, nor on all the worlds which are alike, or on those which are different from these. So that nowhere exists an obstacle to the infinite number of worlds.[3]

Epicurus, without sophisticated observing equipment, had reached much the same conclusion as Sagan and Shklovskii, with years of observational astronomy behind them: the universe is made of pretty much the same "stuff" and behaves similarly no matter where we look.

Indeed, there is plenty of evidence to support the Assumption of Mediocrity. For example, the sun is a very common kind of star. In fact, there are a minimum of one billion sun-like stars in our galaxy alone, and probably many more than that. If the same proportions hold in other galaxies, there may be one hundred billion billion solar-type stars in the universe as a whole.

As for planets, the recent work of Dr. Bruce Campbell and his colleagues is highly encouraging. While the final proof will not be available for a few years, his research does suggest that as many as half of all stars may have large planetary companions.[4] If so, and if a reasonable percentage of them are accompanied by Earth-like planets, then the Assumption of Mediocrity will be given a significant boost.

The basic building blocks of life on Earth also appear to be abundant in the universe. Terrestrial life is built out of just a few elements, such as hydrogen, nitrogen, oxygen, phosphorus, sulphur and carbon. These elements combine together into simple organic molecules that are the foundation of life as we know it. In the past ten years, the work of numerous investigators has revealed that the solar system is awash with "organics."

Simple organic molecules are abundant, not only on Earth, but also in space: "We know now that organic compounds are abundant in everything from the bright tenuous tails of comets to the surfaces of moons and the atmospheres of the giant planets."[5]

It's a long way from simple organics to complex organic molecules, cells, animals, intelligence, and advanced technical civilizations. And yet, if evolution shows us anything, it is a broad tendency to move from simple to complex forms, whether in physics, biology, economics, or sociology. This movement from simplicity to complexity is almost like another law of nature.

Eugene Mallove, author of several books and numerous articles on SETI and related subjects, suggests that matter's evolution into mind is inevitable:

> We cannot prove this now, but as science grows in understanding of matter, chemistry, and life itself, the more apparent it seems that

"life" is the inevitable offspring of senseless matter—given the proper environment. The universe quickens and looks back on itself with a smile. It has come alive.

Magically, quickening means not only literally "coming to life," but also describes the accelerating pace of evolution. Quickening is also the stage of pregnancy in which the movement of the fetus can first be felt. So it seems highly appropriate to suggest that the laws of physics are literally pregnant with life—complexity. They set the stage for life, though not necessarily a specific kind of life.[6]

Mallove is a convincing witness for the case in favor of abundance. Perhaps we do not know the exact mechanism by which life first appeared on Earth, he says, but we do know the laws of physics, and they themselves contain the essential information that will lead to living organisms.

Advocates can buttress their case by calling many credible witnesses to the stand. *Life* Magazine recently polled thirty-five scientists on the question of whether extraterrestrial life exists. Twenty-six were willing to say that life existed in some form in the Milky Way galaxy, four thought there was little chance of it, and five reserved their judgment.

Dr. Frank Drake, a SETI pioneer, current president of the SETI Institute, and an astrophysicist at the University of California at Santa Cruz, said:

We know a great deal about the processes that took place in the solar system and of life on earth, and they were all completely normal processes. No freak events were required. So one would expect that the same sequence has occurred at least in a few places—and probably in many, many places.[7]

Following the same line of reasoning, Dr. Jill Tarter, of NASA's upcoming SETI project, agreed:

The odds are overwhelmingly in favor of intelligent life existing somewhere else in our galaxy. From studying fossil records, there doesn't seem to be anything extraordinarily original about our planet.[8]

Dr. Robert Dixon, director of Ohio State's SETI program ("Big Ear"), said:

Everywhere astronomers look in the universe, we see the same physical laws, the same chemical elements and properties. There is nothing overtly odd about our sun or earth as far as we can discern.[9]

Is the case closed, then? Is it so obvious that life and intelligence are out there, that we simply must search for it because it will be found? If Epicurus, Sagan, Shklovskii, Mallove, Drake, Tarter, Dixon and scores of other experts are right, then the answer to the SETI question is, "No, we are not alone," and the human path of evolution probably leads into a galactic society and all that it entails. Thus, one of the detection experiments now underway *will* confirm that the Assumption of Mediocrity is itself a law of the universe, and we will know the strategy of cosmic evolution.

But wait—the case is by no means closed.

The Case Against

The Assumption of Mediocrity is a convincing argument, but not everyone accepts it. Critics attack it at many different levels, and, at one extreme, there is an alternate school of thought that rejects the most fundamental premises of the case for abundance. The most vociferous dissenters rally around a body of knowledge known as the "Anthropic Cosmological Principle."

To understand this daunting phrase, you must break it down into its constituent parts. The key term is "anthropic," from the same root as anthropology, *anthropos,* a Greek word meaning "man" or "human." Cosmologist Brandon Carter coined the term "Anthropic Principle" in 1974. It is based on the deceptively simple idea that what we see in the universe *must* include the conditions necessary to give rise to intelligent life, or we would not be here to observe those conditions.[10]

It is a simple, almost self-evident statement, and at first, it seems tautological, as does much of the anthropic school's work. The Principle even appears similar to the Assumption of Mediocrity. Both approaches seem to suggest that given the nature of the universe, the evolution of intelligent life is nearly inevitable.

Subtle differences divide the thinking of the two schools. The Anthropic Principle adds something that the Assumption of Mediocrity leaves out, by including the observer in the discussion of the observed. In fact, the existence of "observers" is critical to anthropic thinking, since the presence of observers explains much of the structure of the universe in their model.

The Anthropic Principle also views the presence of many stars and galaxies in the universe quite differently.[11] According to John Barrow and Frank Tipler, two of the strongest advocates of the approach:

> Hubble's classic discovery that the Universe is in a dynamic state of expansion reveals that its size is inextricably bound up with its age. The Universe is fifteen billion light years in size because it is fifteen billion years old.[12]

Astronomer Edwin Hubble discovered that the universe is expanding, with all the galaxies rushing away from each other at the speed of light. This means that if the universe began fifteen billion years ago, it has to be fifteen billion light years in extent. Tipler and Barrow are saying, therefore, that the great size of the universe says nothing about the number of life forms it has produced.

Moreover, the universe could not be much younger or much older than it is, and still have observers in it. The elements from which life is made originate in stars, which then explode into supernovae, scattering those elements out into the universe, so that they can be incorporated into organic structures.

The "cosmic cooking process" within a star takes about ten billion years, so a much younger universe would not yet have produced the elements out of which life is made, and a much older universe might be dying.[13]

To quote Barrow and Tipler again:

> We should not be surprised to observe that the Universe is so large. No astronomer could exist in one that was significantly smaller. The Universe needs to be as big as it is in order to evolve just a single carbon-based life-form.[14]

The same argument has also been advanced by Professor John A. Wheeler, of the University of Texas, another leader in developing the anthropic case. The implications for SETI are described in a 1987 *Discover* article by Dr. Tony Rothman:

> The universe is vast because it's ten billion years old, and it needs to be that old to give rise to one intelligent civilization—which conceivably could be us. Since there's no particular reason why there should be other intelligent beings out there, additional civilizations would just be wasteful of the universe's resources.[15]

Wheeler does not say that we are definitely the only intelligent life form in the universe. Rather, he takes Eugene Mallove's "quickening" argument to a new level. Wheeler might say, "Yes, it does seem that the matter of the universe inevitably gives birth to intelligent life, *but* it is not inevitable that it should happen everywhere. It may well take all the matter, energy, space and time of the universe just to produce what we see on Earth."

Discussion alone will not break this impasse; new facts are needed. If we knew the strategy for the evolution of the universe, we would know whether it required multiple birthing sites for life and intelligence or only one. But it is precisely the strategy that we are trying to discover.

The Anthropic Principle is a rich and complex theory, and no one can do it justice in a few pages. However, doubts have been raised about the Assumption of Mediocrity without delving into the complexities of anthropic thinking.

Other critics of the case for an abundance of life and intelligence say, "It's not at all inevitable that there's an abundance of intelligent life in the universe just because there are billions of stars, or lots of planets. The evolution of life on Earth, and especially intelligent life, depended on a number of highly improbable occurrences. If any one of those factors were lacking on another planet, neither life nor intelligence would appear."

Gerald Hawkins, who is optimistic that the next great "mindstep" in human evolution will be contact with extraterrestrial intelligence, is still aware of the problems that are involved. He quotes pathologist Leonard Ornstein, of the City University of New York, who argues that the evolutionary track leading to intelligence is "fraught with chance and risk." According to Ornstein, "the dice have been thrown in such a bizarre sequence, it could happen once, and once only."[16]

Ornstein contends, in particular, that the probability argument underpinning the abundance case does not hold:

> I have a billion marbles in a bag. The first one I pick out is labeled "a planet with life." What are the chances of picking out a twin? Strict probability says there may be no twin left behind in the bag. All the other marbles may turn out to be labeled "a planet with no life." So earthlings can't prove (or disprove) there are other earthlings out there from the single labeled marble. It is a matter of believing or disbelieving.[17]

Advocates of the case for scarcity can turn the argument from the numbers against their opponents in other ways as well, especially with a challenge known as "Fermi's Paradox." In the middle of a discussion with his colleagues one day, the great physicist Enrico Fermi said simply, "Where are they?" He meant, "Where are the extraterrestrials?" because if life and intelligence are so abundant in the universe, then why haven't we heard from them or seen them yet?

As Professor George Field of the Smithsonian Center for Astrophysics at Harvard University points out, speculation on SETI today can be analyzed according to how people deal with "Fermi's Paradox." There are three basic responses:

1. Some say, "They are here," pointing to UFO sightings and arguing that these are extraterrestrial spaceships
2. Some say, "They are there," embracing the Assumption of Mediocrity, and urging increased efforts to detect the signals that extraterrestrial civilizations might be sending to us
3. Some say, "They are nowhere," arguing that intelligent life on Earth is a highly contingent occurrence, not necessarily repeated anywhere else[18]

Frank Tipler believes that the failure of SETI enthusiasts to resolve the Fermi Paradox shows that intelligent extraterrestrials simply aren't present, at least in our galaxy. The Anthropic Principle aside, Tipler thinks the belief in extraterrestrials is based on shaky assumptions:

> The contemporary advocates for the existence of such beings seem to be primarily astronomers and physicists . . . while most leading

experts in evolutionary biology . . . contend that the earth is probably unique in harboring intelligence, at least among the planets of our galaxy. I agree with the biologists . . .[19]

Tipler bases his opinion on one simple proposition: they aren't here, and therefore they don't exist. He extrapolates from the observed tendency of terrestrial intelligence to send out exploratory probes as soon as the technology is available. He goes one step further and imagines that such probes can become semi-autonomous, using the resources of other star systems to construct new probes, which would in turn begin their own exploration.[20]

Estimating 100,000 years per interstellar flight and 1,000 years to build a new probe on arrival in a new star system, he argues that a single probe and its descendants would have explored the entire galaxy in 300 million years. Based on Earth's history, Tipler estimates that it would take six billion years from the formation of a planet until an intelligent species had begun sending out probes. Thus, we should have heard by now from any star system over 6.3 billion years old (six billion plus 300 million), and over half the stars in the galaxy are at least that old.[21]

Ironically, those who believe that UFOS are extraterrestrial spaceships would find Tipler's argument supportive of their own. They would only disagree with his statement that "they don't exist because they aren't here."

However, most SETI scientists do not find the UFO argument convincing, and they therefore must agree with Tipler that the extraterrestrials have not yet arrived. Having made that admission, however, they are left with only the most speculative arguments about why this would be so.

Interstellar migration analyses provide another argument similar to Tipler's discussion of probes. According to Dr. Eric Jones, an expert on the topic, a civilization able to control the resources of an entire star system should be able to move out quite rapidly into the galaxy. Jones estimates that colonization of the entire galaxy could occur in about 100 million years, a third of the time Tipler estimates that it would take to explore the galaxy with unmanned probes.[22]

Jones' analysis is a variation on the question of "Why aren't they here?" In addition, a question arises out of SETI itself. If they exist, why haven't we heard anything in thirty years of listening? If the galaxy is teeming with advanced civilizations, it's difficult to imagine why the SETI efforts already undertaken haven't produced more results.

Finally, there is a critique of the case for abundance based on one of its strongest arguments: the concept of a uniform evolutionary process throughout the universe. In response to the *Life* poll on the probability of life and intelligence existing in the universe, a number of scientists focused on that aspect of the problem. For example, Gerrit Verschuur, a radio astronomer, said:

> Intelligent life like us? Practically zero. But I believe that life is common in the galaxy. There are an incredible number of inhabited planets—with some form of primeval life—but the number of planets in our galaxy on which life-forms will be like us is incredibly small.[23]

Eric Chaisson, senior scientist at Baltimore's Space Telescope Science Institute, says that there are two major hurdles to overcome before advanced communicating civilizations can come into existence. The first barrier is the step from inorganic to organic matter. The second is from organic matter to intelligence:

> While the step from inanimate matter to the origins of life seems relatively smooth, the great step beyond . . . is a much bigger jump.[24]

Edward Olson, an astronomer at the University of Illinois agrees, suggesting that the likelihood of lower forms of life being prevalent are "almost one hundred percent" but the chances of intelligent life are small. He argues that the evolutionary path is not a smooth or predictable one:

> The whole history of life on earth seems to have been composed of periods during which new species appeared very rapidly. At other times, there were mass extinctions. The whole process seems to have proceeded in a jumpy way. . . . Astronomers who are optimistic about intelligent life elsewhere tend to think of evolution as a linear process. But it isn't . . .[25]

The argument that evolution is "the same" everywhere in the universe may not mean much. Unless it is *exactly* the same as on

Earth, it may not produce life, or it may produce life, and not intelligence. Mars appears to have been moving toward living organisms and perhaps even intelligent beings, but it stopped for reasons we do not yet understand. Perhaps there were many possible paths for Earth to have taken, any one of which might not have produced life, or us.

The distinction between the evolution of life and the emergence of intelligent life is rarely made by those who aren't deeply immersed in the debate. However, the distinction is critical, and those who believe life is abundant but intelligence is rare are emerging as a third school of thought.

Even the most ardent proponent of the abundance position must feel a bit uncertain in the face of these counter-arguments. There are those who begin with a belief in the Assumption of Mediocrity, only to abandon it after further investigation. Eric Jones once had a deep fascination with SETI, only to decide, based on his own research, that there were better places to put his energy. Gerrit Verschuur took both SETI and UFOs seriously for a time, but then concluded that neither was worth further investigation.[26]

However, neither side of the debate has presented the final answer to the extraterrestrial question, yet. Abundance advocates do not give in easily, and they have answers for their critics.

Rebuttals

To the argument that the resources of an entire universe or galaxy might be needed to create one intelligent species, author Isaac Asimov says:

> You might need an extremely large universe to supply the kind of properties that would make it possible for life to develop on one world. But having achieved that, it doesn't mean that life will develop only on one world. In order for a raindrop to form, you have to have some concatenation of clouds and wind patterns and temperature change that will involve a large section of the Earth. But one raindrop isn't the only one that forms; billions of raindrops form, and I think if the universe had to be large enough for one world to develop life, it might well mean that many worlds could then develop.[27]

Dr. John Billingham, now directing the SETI project for NASA, agrees:

To me, it is perfectly logical that life began early in the history of Earth, and I don't see why it wouldn't happen elsewhere. Because it *has* occurred is, to me, actually strong evidence that it is a natural phenomenon.[28]

In response to why the extraterrestrials haven't revealed themselves to us, Dr. Michael Papagiannis of Boston University, one of the founders of the field of bioastronomy, argues that they may be observing us and waiting until we are ready for the shock of contact. He thinks that if a galactic civilization exists, it would have had experience in bringing less advanced cultures into the wider society, and they may know that the process cannot be rushed.[29]

Author Rick Cook argues that the model of space settlement developed by Dr. Gerard K. O'Neill can be an answer to the Fermi Paradox. He says that a civilization intent upon spreading through the galaxy could do so quickly. However, once a civilization had begun to use the resources of a solar system, it might slow its own expansion:

The O'Neill paradigm may slow the development or expansion of a species through the galaxy. In this model, you have billions of people living in tens of thousands of space habitats, and people will become adapted to that environment. They may not necessarily keep moving, and a spacefaring civilization may become more static in time.

You don't have to jump to the next Earth-like planet nor necessarily even to the next star. If you don't like your neighbors, you just start a new colony . . . this may slow the immigration process.[30]

Dr. Bernard Oliver of the NASA SETI project goes one step further, arguing that the return on the enormous investment in energy and resources for interstellar travel is too small to justify it. In his view, the commercial motivation for colonization is not there, and so the extraterrestrials are ". . . home minding their own business because they know how expensive it is to get here."[31]

Finally, Kent Cullers of the NASA project says that we can't draw any significant conclusions from the lack of success with listening efforts to date, because the "search space" is so large, and the technology used has been so limited. "In effect," he says, "all the searches to date would not have found another civilization like the Earth even orbiting the nearest star."[32]

The debate will continue, as it must, until we have new observational evidence supporting one side or the other. In the meantime, most scientists, regardless of where they stand on the issue, think that SETI is worthwhile. That is partly because of its relatively low cost, partly because of its potentially great return, and partly because it is a scientific experiment that will generate new knowledge about our universe, no matter what the results are.

Most would agree, then, with Dr. Philip Morrison of MIT, whose paper (with Cocconi) helped start it all back in 1959: "I don't think it does any good to sit here and place bets. I'm not in favor of guessing, I'm in favor of looking."[33]

Dr. Morrison's approach represents a fourth school of thought, which is the empirical school. The empiricists view the debate as somewhat interesting intellectually, but not worth a lot of time and energy. For the first time in history, humans have the technology to settle the question, so let's just get on with it.

And we are. The search is on and the experiments are underway. Unlike a real courtroom, where the jury must make its decision based on a limited amount of evidence, there is still much more evidence to be brought in.

Whatever the result, it will have a profound impact on our society because it speaks to something very deep within us. Even if contact never occurs, we learn from the search. It is focused externally, but reflects deep internal needs on the part of human beings. SETI is an effort to determine if we are alone in the universe, and it reflects a hope that we are not. We want to find someone or something "out there" to help us better understand our own mysterious natures.

While the search is stated in scientific and social terms, it has much the same psychological structure as a religious search for God or a Higher Power. As David Williamson, Special Assistant for Policy Integration at NASA, has said, space exploration is about hope, a commodity that the human species needs badly. By supporting hope, the search creates its own value, regardless of what is found.[34]

The SETI Factor is a continuing force in human affairs. For some, it is a belief that wise and benign extraterrestrials will save humanity from itself and show us a better way of life. For others, thinking about extraterrestrials better defines what it means to be a human being.

SETI won't let us escape the issue of identity. As the greatest human philosophers and religious teachers have taught us, we cannot understand ourselves except in contrast with others. "I know who I am because I'm different from you," or "We know we're Americans because we're different from the Russians." The SETI Factor, as it has evolved in modern times, brings the definition of identity to a new, planetary level: "I know what it means to be an Earthling because it's different from being a Tau Cetan or a Vegan."

Thinking about the promise and perils of SETI creates a feeling of anticipation and uncertainty. A valid detection might happen tomorrow, or next year, in 1992, or in the year 2000. Is humanity "ready for SETI"? Perhaps not, but getting ready may be humanity's most important task in preparing for the Third Millennium.

Chapter Four

The SETI Factor

> The question of extraterrestrial life, rather than having
> arisen in the twentieth century, has been debated almost
> from the beginning of recorded history. Between the fifth-
> century BC flowering of Greek civilization and 1917,
> more than 140 books and thousands of essays, reviews,
> and other writings had been devoted to discussing
> whether or not other inhabited worlds exist in the
> universe. Moreover, . . . the majority of educated people
> since around 1700 have accepted the theory of
> extraterrestrial life and in numerous instances have
> formulated their philosophical and religious positions in
> relation to it. To put this point differently, even if no
> UFOs hover in our heavens, belief in extraterrestrial
> beings has hovered in the human consciousness.
>
> —Michael J. Crowe, *The Extraterrestrial*
> *Life Debate 1750–1900*

THE SETI FACTOR HAS EXERTED A STRONG INFLUENCE ON HUMAN
society in the past, and it will continue to do so in the future.
As the search continues, its results will not be fed into a vacuum.
People and societies with pre-existing ways of looking at everything,
including extraterrestrials, will process the information gathered by
the searchers. These cultural filters are an important variable determining the impact of SETI. In this chapter, we consider the historical
perspective, laying the foundations for discussion of future impact.

According to a study of impact by NASA sociologist Dr. Mary
Connors, contact will be perceived by social groups through their
existing "cognitive filters" or worldviews.[1] Contact will mean one
thing to the SETI scientist and another to the theologian. It will

affect a citizen of the United States according to American attitudes and an Australian aboriginal according to that culture's own cosmologies. Christians will see it through their theological filters, while Muslims will perceive it according to theirs.

Belief systems are ways of viewing reality. Without our being conscious of it, they create reality. However, belief systems are not merely mental constructions, they also serve powerful social functions. In the language of evolutionary psychology, belief systems are "adaptive." The person who sees the world through a given system is convinced that it will contribute to his or her survival.

Dr. Richard Robbins, an anthropologist and expert on belief systems, says that the function of a belief system for a social group is to maintain its power and status in a society. How a group interprets any new information depends on whether it will enhance or diminish their status.[2] The Catholic Church's rejection of the Copernican Principle is a good example of the behavior Robbins describes.

Robbins suggests that the reaction of social groups on Earth to news of contact will depend upon the perceived effect on their social status. His observations tend to confirm comments made by science writer Andrew Chaikin, who points out that the desire to understand the meaning of the signal and the fierce media attention on the story would suddenly make astronomers among the most powerful people in the world. Their power will flow from the fact that they have the expertise to interpret the context and relevance of the signal. The social status of astronomers will be enhanced by contact.[3]

The modern SETI movement is only thirty years old. However, once the search is seen as part of the psychological makeup of human beings, today's scientific search can be seen as continuing something that began long ago, but was described in different language.

The scientific method offers a different approach to verifying the existence of extraterrestrial intelligence, but differences in methods and language should not obscure the common threads that have persisted over time.

Today, the search for extraterrestrial intelligence is conducted by scientists and described in scientific terms. We are so accustomed to this approach that we take it for granted, assuming that it is different from a search for God, or philosophical speculation about the nature of the universe. However, the impulses of the scientist, philosopher, and spiritual seeker may be the same. All are displaying an intense desire to know more about ourselves and the universe.

Chet Raymo, columnist for the Boston *Globe* analyzed this parallel when he compared the work of cosmologist Stephen Hawking and actress/New Age guru Shirley MacLaine:

> What the vivacious guru and the professor have in common is that they both sell lots of books. . . . Apparently, both authors have something to say that the public wants to hear.
>   Both assert the same goal—to know how the universe works and the role we play in it.[4]

If contact is made with extraterrestrial civilizations, the reaction of human society will not be limited to the scientific community. The response will emanate from many different peoples and cultures with contrasting belief systems. We should understand that these psychological and sociological structures provide the main foundation for understanding the potential impact of contact. On the other hand, if there are psychological patterns and structures *common* to all human cultures, then that would be a counterweight to the diversity of response, and would provide another avenue to understanding impact.

There are common perspectives, and many of them have been uncovered by scholars such as Joseph Campbell, who discovered a multitude of commonalities when he analyzed the myths of different world cultures. The psychologist Carl Jung also found universal archetypes that he considered part of the "collective unconscious" of humanity.

From the perspective of SETI, one of the most important common themes found in human cultures is the idea that life and intelligence on Earth have an extraterrestrial origin.

Creation Myths: "Ourselves in the Universe"

A myth is a story that people tell in order to reveal or explain something important about themselves. It usually has a valid historical basis, but it may not be wholly factual. Accuracy is, however, far less important than the myth's value in revealing other kinds of truths.[5]

For example, there are many creation myths surrounding the birth of the United States of America, especially concerning the "Father of the Country," George Washington. Did George Washington throw a coin across the Potomac River, and did he tell his father the truth about chopping down the cherry tree?

Myths are more about social psychology than history, even though they are often historically based. Seen in that light, they offer revealing glimpses into a culture's deepest concerns: Throwing the coin across the river says, "George Washington was strong," and it establishes a role model for a people living on the frontier. They were faced with the challenge of settling the wilderness, and with creating a new nation. The people of the United States needed to be strong to do that.

The message of the cherry tree episode is, "George Washington was honest." The new nation needed that type of moral fortitude as it struggled to create a country standing in sharp contrast to the political systems it left behind.

The number of creation myths with an extraterrestrial dimension is a striking feature of terrestrial cultures—a common theme in the midst of great diversity. These myths sometimes imply that humans actually came here from other planets, and often assert that culture-bearers arrived from space to enlighten a backward human species.

The belief in the power and importance of these "sky-gods" is not universal. It competes with an earlier worship of terrestrial gods and goddesses, and conflict between the two belief systems is yet another theme of human history. The extraterrestrial connection seems surprising only because the Darwinian theory of the purely terrestrial origin of life has become so dominant in the past hundred years. Darwinism has supplanted many of the creation myths, even though it has not been proven beyond a reasonable doubt.

Let's take a brief look at a few creation stories to get a flavor of how humanity views itself in relation to the universe.

For those raised in the Judeo-Christian tradition, the description of creation given in Genesis is the most familiar:

In the beginning, God created the heavens and the earth. The earth was without form, and void; and darkness was on the face of the deep. And the Spirit of God was hovering over the face of the waters.[6]

In this description, God creates by making distinctions. First, He separates light from darkness and day from night. He then separates Heaven from Earth:

Then God said, "let there be a firmament in the midst of the waters, and let it divide the waters from the waters. Thus, God made the firmament, and divided the waters which were under the firmament from the waters which were above the firmament. . . . And God called the firmament Heaven. . . . Then God said, "Let the waters under the heavens be gathered into one place, and let the dry land appear. . . . And God called the dry land Earth. . . .[7]

The origin of life is the result of the divine force at work. On successive days of creation, God brings forth the grasses, herbs, and seed plants; water creatures and birds; land animals of all kinds; and, finally, humans.

The origin of intelligence is a bit more problematic. Adam and Eve are far beyond the other animals from the very beginning. However, *self-awareness* seems to arise through the eating of the fruit of the Tree of Knowledge of Good and Evil, an action brought about by the intervention of Satan in the form of a serpent. Genesis appears to say that life is good in itself, but intelligence and self-awareness can be used for good or evil—an idea that would find many adherents today.

One of several Greek creation myths tells a story that is both similar and different:

In the beginning, Eurynome, the Goddess of All Things, rose naked from Chaos, but found nothing substantial for her feet to rest upon, and therefore divided the sea from the sky, dancing lonely upon its waves.[8]

Eurynome couples with the great serpent Ophion, then assumes the form of a dove, and lays the "Universal Egg." It hatches and splits in two:

Out tumbled all things that exist, her children; sun, moon, planets, stars, the earth with its mountains and rivers, its trees, herbs, and living creatures.[9]

Metaphorically, this description is not far from the "Big Bang," in which all the matter and energy of the universe emerge from a primeval atom.

Eurynome and Ophion retire to live on Mount Olympus, but she becomes angry with him when he claims all the credit for being the sole creator of the universe. She kicks him in the head and banishes him to the caverns under the Earth.[10]

In both descriptions of creation, the separation of Heaven and Earth is central, as is the conflict with the serpent, who seems to embody our capacity for evil.

Islam, which shares many traditions with the Judeo-Christian religions, also declares God as the creator of Heaven, Earth, human beings, and all other living things.[11]

The theory of evolution and the Assumption of Mediocrity contradict two of humanity's strongest impulses, which are to think of ourselves as special and to make a distinction between "Heaven" and "Earth."

From the earliest days, humans have considered themselves unique because of the interest that God, or the gods, have shown for our planet. In the many spiritual traditions, God is seen as not only creating the earth, but also remaining involved with its inhabitants, establishing standards of ethics and morality for them.

The separation of Heaven and Earth is especially visible in religious thought and language; however, it is present in secular thinking as well. The physical separation of Heaven and Earth mirrors a psychological separation that is even more important: Heaven is the home of the Creator, and Earth the home of the created. The separation is also linked to humanity's shortcomings. In the Bible, the most traumatic split between Heaven and Earth occurs not with the Creation but with the Fall, when Adam and Eve disobey God and eat the forbidden fruit. They are driven from the Garden of Eden, and angels are set at the entrance to prevent their return.

Before the Fall, God walks in Eden and has contact with his creatures. Afterward, the relationship changes, and both God and

Heaven become distant and unattainable. Later, when humans try to reach Heaven without permission, as at the Tower of Babel, they are punished. This description is also consistent with numerous other cultures' memories of a "Golden Age," when the gods walked on Earth and all was well, followed by a period in which the gods have gone away.[12]

In Greek mythology, the peak of Mount Olympus represents the "home of the gods," from which they looked down on human affairs. The gods frequently traveled to Earth, intervening in human conflicts, such as the Trojan War, but humans did not visit Mount Olympus without suffering grievous consequences. In Norse mythology, warriors who died in battle were said to be carried away to Valhalla, a boisterous Scandinavian version of Heaven.

Christian thought maintains the distinction between Heaven and Earth. The Lord's Prayer begins with "Our Father, who art in Heaven." Jesus often reminds his followers that he has come down from Heaven to be with them, and that his kingdom is "not of this world."[13] In many of Earth's religions, Heaven is good, and Earth is bad, and the solution to our problems is seen as coming from Heaven.

There are other beliefs, to be sure. In Eastern religions such as Buddhism and Hinduism, the earth is a plane of existence on which human beings work through the consequences of their deeds and misdeeds of past lives. As they evolve spiritually, they reach various heavens that are far more pleasant places. The earth is not bad; it is a domain in which our "karma" is worked out and we are released to participate in higher levels of existence.

In religions that focus on "goddess" imagery, the "Earth Mother" is revered, worshipped. Tremendous love for the earth can be seen in the spiritual life of the Native American culture. Many religions also downplay the idea of salvation coming from outside ourselves, including certain aspects of Christianity. (Jesus said that "the Kingdom of Heaven is within.")[14] Buddhism does not promote the idea of an omniscient creator-God who controls our lives or saves us from sin.

The great transformations in human thought that began with the Renaissance have loosened the hold of earlier perceptions on the conscious mind. To the twentieth-century person, "Heaven" may be a metaphor, not a physical place, and the concept of an earth created by chance and evolution is commonly accepted.

However, the structure of the beliefs remains in place, even though the content has changed. The terms "outer space," or "the universe" are used rather than "Heaven" or "the heavens." For many, however, there remains a sense of awe and mystery that is close to religious in its expression.

Even today, there is a projection of positive values onto the extraterrestrial dimension. For example, "before and after" thinking characterizes the pro-space movement, which is dedicated to supporting a migration of humanity off the home planet. The "before" phase is that period of time in which humanity has been confined to Earth as a "single-planet" species. The "after" phase comes when we break the bonds of gravity and spread outward into the universe as a "multi-planet" species.[15]

Will humanity's lot improve as it becomes a spacefaring culture? That can only be determined once the step is taken, but the commonality of ancient and modern thinking remains in place.

Among other social groups, many New Age thinkers have become interested in the UFO phenomenon, and believe that extraterrestrials will help us avoid nuclear war or ecological catastrophe, and transform our planet.[16]

Thus, the belief in salvation from an extraterrestrial source remains intact, even though it is expressed quite differently than in the past.

As Andrew Chaikin puts it:

> It is part of the feeling of being confronted with an entity from all that "other" that we know so little about. We imbue this entity with knowledge far beyond what we know, out of a wish that the ET's will be saviours, or god-like in some way.[17]

The expected benefit of contacting an advanced extraterrestrial civilization is also an explicit element of the scientific search. Terrestrial civilization has reached an uncertain phase of development and our technological culture has created the possibility of extinction for our species. We are groping for solutions, and it would be comforting to talk with a civilization that has passed this stage, knows that it can be survived, and how to do it.

As Dr. Jill Tarter puts it:

> If we detect a signal, the only way that can happen is if technological civilizations, on the average, lived for a long time. Just the detection of a signal, with no information content, tells us that it is possible to survive our technological infancy, the state we're in now.[18]

If SETI produces that simple result, it will be revolutionary. Critics of advanced technological society believe sincerely that humanity is going in the wrong direction, and there is plenty of evidence that they are right.[19] The only way of determining whether it is the wrong path without going down it is to discover that other cultures made it through their technological adolescence, and to discover the causes of their success.

There are many reasons for humans' intense interest in outer space and what it contains, and some of them fit easily into the modern understanding of life and evolution, without contradicting the earlier cultural models.

For example, there are many people who believe that the gods and goddesses of ancient times were extraterrestrials. Perhaps they came to Earth on exploratory missions, and because they were so far advanced beyond the cultures of the times they were perceived as supernatural beings.

The evidence for this remains inconclusive, but it is not illogical, and it is an answer to Fermi's Paradox. Human beings plan to explore the solar system and the galaxy, and are likely to do so eventually. If evolution is similar everywhere in the universe, then there will be planets at much earlier stages of development than our own. If a twenty-second century spacecraft from Earth discovers a planet at the level of Greece in the fifth century BC, the local culture might well enshrine the landing party as gods.

Many scientific models of life's origins also do not require a purely terrestrial origin. For example, comets and meteorites seem to be rich in the organics that are the precursors for life. Meteorites hit the earth, and the planet passes through the tails of comets. These encounters offer the possibility of Earth being "infected" with life-producing materials. That's a far cry from a creator-God consciously creating the universe, but it *is* a creation story that involves an interaction of Heaven and Earth.

What is more important than the *truth*, however, is the wide-spread *belief* in an extraterrestrial origin, involvement, and/or destiny for humanity. To the extent that this is a common assumption (conscious or unconscious), it will play a major role in determining how human beings respond to contact, or a lack of it.

Rethinking the Distinction

Whether the matter is seen in terms of physical or metaphorical reality, God is, by definition, an extraterrestrial force, as are many of the gods. For many believers, God exists not only off of the planet Earth, but also outside of human time and history. God is seen, however, as intervening in human history and there are attempts to bring Heaven and Earth closer together through divine intervention.

Language and culture are critical factors in thinking about gods in this way. In the centuries before Jesus, people thought they saw beings who came to and went from Earth and called them "the gods." Later, perhaps they were called angels. Today, people may see the same things and call them extraterrestrials.

According to some accounts, cosmonauts on the Soviet space station recently reported that they had seen a group of seven "angels" outside the station. The mission controllers on Earth did not take this report too seriously until they sent up a relief crew that also saw the "angels."[20] Very little has been said about this incident since it occurred, but one tabloid paper reported that a Soviet academician had investigated, and came back with reassuring news: the apparitions were not angels, he said, but beings that had evolved to a higher energy level and no longer needed bodies![21]

Who knows what really happened on the Soviet space station? The entire episode is bizarre, but the explanations show how different cultures respond to such events. Apparently, the Soviet authorities could not accept an explanation with religious connotations. However, the explanation that did suffice would never have been made by NASA because then it would posit the existence of something equally strange.

Cultural differences will persist in the future, but the SETI Factor reveals a common myth tradition in which life and intelligence

flower on Earth because of extraterrestrial intervention, and the
future of life and intelligence on Earth are insured by future contact.

Seeking salvation from outside ourselves is clearly part of the
"extraterrestrial connection," it is an element of the search, and it is
important to understand how it is helping to drive the SETI effort.
However, it is not the only motivation. Curiosity, for example, is
another human trait almost as powerful as the seeking of a Higher
Power. Exploration, the result of curiosity, is also a behavior pattern
deeply embedded in the human psyche.

SETI is an effort at self-definition, which is just as important to
human beings as seeking salvation from external forces, and can
serve as a counterbalance to it. It may be that we cannot be "saved"
by an external force, that our salvation must be worked out by
ourselves "with fear and trembling." By thinking about how other
forms of intelligent life have evolved, human beings create entirely
new views of themselves, such as the Assumption of Mediocrity and
the Anthropic Cosmological Principle.

For all that humans have learned about themselves, we still do not
know whether we are a unique or common life form. The search for
extraterrestrial intelligence is therefore a search for personal and
planetary identity.

Contact Myths

It is legitimate, but difficult, to look for evidence of extraterrestrial
intelligence having already visited Earth. In their book, Shklovskii
and Sagan speculate not only on the probability that intelligent
extraterrestrial life exists, but also that direct encounters between
extraterrestrials and Earthlings might have occurred in the past.

How, the authors ask, would we have known about these encounters if they had taken place? The answer is that we would find them
in the myths that cultures tell about their own history. Shklovskii
and Sagan found a plausible "contact myth" in some of the stories
surrounding the origins of Sumerian civilization, an early human
culture that began in the fourth millennium BC in what is now Iran.

According to Sagan and Shklovskii, the original contact myth of
the Sumerians evolved into a religion that is almost totally oriented
to the stars:

The gods are characterized by a variety of forms, not all human. They are celestial in origin. In general, each is associated with a different star.[22]

The legends come to us from the time of Alexander the Great, and tell how the Sumerians originally lived "without rule or order, like the beasts of the field." Then, a great change occurred, when "an animal endowed with reason," named Oannes appeared out of the Persian Gulf area.[23]

Oannes is described as a being of striking appearance:

The whole body of the animal was like that of a fish; and had under the fish's head another head, and also feet below, similar to those of a man, subjoined to the fish's tail. His voice, too, and language were articulate and human. . . .[24]

Oannes brought civilization to the Sumerians, teaching them everything they needed to know in order to take the next steps in social evolution:

He gave them insight into letters, and sciences, and every kind of art. He taught them to construct houses, to found temples, to compile laws, and explained to them the principles of geometric knowledge. He made them distinguish the seeds of the earth, and showed them how to collect fruits. In short, he instructed them in everything that could tend to soften manners and humanise mankind. From that time, so universal were his instructions, nothing material has been added in the way of improvement.[25]

Four other teachers followed Oannes at various times in Sumerian history; they all looked like Oannes and came out of the Persian Gulf area.

The Sumerian story is the only one cited by Sagan and Shklovskii as being a potential contact myth. However, similar patterns hold in widely divergent cultures: there is a bringer of culture, who either comes to a society from the stars, or from an unknown place far away. The gods and humans often interact extensively; sometimes semi-divine children are born on Earth.[26]

Even the Judeo-Christian description of the development of human civilization includes such passages, as in Genesis:

Now it came to pass, when men began to multiply on the face of the earth, and daughters were born to them, that the sons of God saw the daughters of men, that they were beautiful; and they took wives for themselves of all whom they chose.

There were giants on the earth in those days, and also afterward, when the sons of God came in to the daughters of men and they bore children to them. These were the mighty men of old, men of renown.[27]

The Genesis description resembles Greek myths that include detailed descriptions of the gods and their interactions with humans. The Greek legends tell many stories of gods and humans falling in love and having children who are "men of renown," such as Hercules and Achilles.

Those stories are also consistent with the myth of Quetzlcoatl, who became a god of the Aztec Indian culture. Quetzlcoatl apparently took his name from the quetzal bird, which has brilliant green and red plumage. He was also known as the "feathered serpent," because he had plumage running from his head down his back.

Quetzlcoatl appeared one day to the people and remained among them for many years, teaching them the ways of civilization, much as Oannes had done for the Sumerians. He lost his power and prestige because he became intimately involved with human women, which he considered to be a sin. He then went back "into the sky," but promised he would return.

Quetzlcoatl was different from the native peoples in that he was light-skinned, and that may be partially responsible for the destruction of the Indian civilizations of Mexico. Believing that their god would return, the Indians were looking for benign, light-skinned "gods," and mistook the Spaniards for their saviors. The result was a disaster for them, as civilizations consisting of thousands of people were defeated by a few soldiers and adventurers.

In a commentary on the book of wisdom known as the *I Ching*, there is a description of the founding of Chinese civilization by one Fu Hsi:

In the beginning there was as yet no moral or social order. Men knew their mothers only, not their fathers. When hungry, they searched for food; when satisfied, they threw away the remnants. They devoured their food hide and hair, drank the blood and clad themselves in skins and rushes. Then came Fu Hsi and looked upward and contemplated the images in the heavens, and looked downward and contemplated the occurrences on earth. He united man and wife, regulated the five stages of change, and laid down the laws of humanity. . . .[28]

All of these stories can be understood as simple ways of explaining the complex evolution of human societies. Rather than transforming because of the heroic intervention of one person, societies may evolve naturally from one stage to another because of the collective activities of all their members. The mythical personages can be representations of entire cultures at a given time.

Still, the contact myths must be taken into account in thinking about contact with extraterrestrials in the future, because much of the Earth's population once believed or still believes that human beings and/or human civilization are extraterrestrial in origin.[29]

The Continuing Contact Myth

The contact myth did not die in Sumeria or Mexico; it continued to be told and experienced in medieval Europe and persists even today.

Jacques Vallee is an information scientist who has conducted extensive research into the UFO phenomenon of the past and the present; he was the model for the French scientist who led the research team in *Close Encounters of the Third Kind*. Vallee understands the scientific reluctance to investigate a phenomenon that offers no physical evidence for examination. He argues that there are artifacts that can be investigated scientifically, and these are the *reports* of UFO sightings.

His review of the historical record shows that some variation of this phenomenon has existed since the dawn of civilization. In 1965, Vallee reported that he had files on some three hundred sightings prior to the twentieth century, sixty from the years before 1800.

Quoting another researcher, Vallee describes one of these incidents:

Agobard, Archbishop of Lyons, wrote in *De Grandine et Tonitrua* how in 840 AD he found the mob in Lyons lynching three men and a woman accused of landing from a cloudship from the aerial region of Magonia. The great German philosopher, Jacob Grimm, about 1820 described the legend of a ship from the clouds, and Montanus, the eighteenth century writer on folklore, told of wizards flying in the clouds, who were shot down. The belief of beings from the skies who surveyed our Earth persisted in human consciousness throughout the Middle Ages.[30]

Vallee confirms that people did not talk about spaceships or extraterrestrials in Greece, Rome, or medieval Europe:

> The appearances of lights or phenomena interpreted as objects, seen in the sky, were not in general associated with the idea of "visitors" or with the possible arrival of fantastic creatures, but rather with religious beliefs and were treated as manifestations of supernatural forces.[31]

Today, the tradition continues with "contactees" and "abductees," some of whom claim to have been taken aboard extraterrestrial spaceships and even impregnated by aliens.[32] Debate is raging over whether their experiences are valid, but that may not be too important in the long run. These experiences continue in the tradition of those who met Oannes and Quetzlcoatl, and dissemination of the stories has an effect, as did those earlier stories. In the words of Jacques Vallee:

> In a particular society if enough people believe in something, then that something exists. To paraphrase one of the founders of modern sociology: 'if men believe something to be real, then it is real in its consequences.'[33]

Michael J. Crowe says, ". . . even if no extraterrestrials exist, their influence on terrestrials has been immense."[34]

This brief review of human thinking reveals that the extraterrestrial dimension has already exerted a tremendous influence on society. When the idea of an extraterrestrial is broadened, the SETI Factor is revealed as woven into our most important ideas about human origins and destiny.

As modern SETI begins to dominate the dialogue, major changes in human perceptions of extraterrestrials may be anticipated in the very near future.

Chapter Five

Modern SETI

Starting with Project Ozma in 1959, there have been some forty-nine separate radio searches for ETI signals. . . .

—"SETI: The Search for Extraterrestrial Intelligence Program Plan," NASA

A New Form of Space Exploration

IN THEIR 1959 PAPER, "SEARCHING FOR INTERSTELLAR COMMUNICATIONS," Giuseppe Cocconi and Philip Morrison suggested that the time had come to begin searching for electromagnetic signals that might have been sent by extraterrestrial civilizations.

We now view the Cocconi/Morrison paper as the catalyst for modern SETI, perhaps because it appeared at a propitious time. The technology existed in radio telescopes to carry out their suggestions and that fact alone made the suggestion intriguing. That two well-known and highly-respected physicists made the proposal gave the idea a validity it might not have claimed otherwise. In addition, the scientific exploration of outer space had begun some two years before, with the launching of *Sputnik* by the Soviet Union, generating new enthusiasm for space-related ventures.

Beyond the basic arguments about the probable existence of intelligent life, the authors proposed that if extraterrestrials were communicating, then electromagnetic radiation would be the best method over interstellar distances. These signals travel at the speed of light, and at certain frequencies are not easily absorbed by planetary atmospheres or the interstellar medium (the space between the stars.)[1]

They further suggested that it made sense to listen at the radio frequency of 1420 megahertz, and a corresponding wavelength of 21 centimeters. (For comparison, a frequency of one megacycle per second, or one megahertz, is the frequency in the middle of the AM radio band.)

Neutral hydrogen is the most abundant element in the universe, and the frequency/wavelength combination recommended by Cocconi and Morrison is the location of natural emission for neutral hydrogen. They reasoned, therefore, that it was an "objective standard of frequency, which must be known to every observer in the universe."[2]

Cocconi and Morrison reached their conclusion based on their understanding of the barriers to radio transmission within the galaxy, and their insight as to how those barriers might be overcome. The interstellar medium is a major problem, as are the atmospheres of planets. In addition, all the matter in the universe is radiating energy, setting up a kind of "cosmic background noise" that would interfere with any artificial signals.

These different barriers make it hard to send a signal randomly because it will most likely be absorbed along the way. Calculations showed that the best wavelength range for receiving signals is quite restricted, between 3 and 30 centimeters.[3] In addition, it would be almost impossible to pick up a signal if the receiving civilization did not know the transmission frequency in advance. Sagan and Shklovskii calculate that there are still nine billion potential transmission frequencies between 3 and 30 centimeters! As they point out, multiplying the number of possible frequencies by the number of possible inhabited planets makes the problem of guessing the right frequency a serious one.[4]

Sagan and Shklovskii explain that the neutral line of hydrogen might be a universal clue that would tell communicating civilizations where to "turn the dial" to be in touch:

[Cocconi and Morrison] arrived at the elegant idea that nature itself provides a standard calibrating frequency within this wavelength range—namely the 21 cm (1420 megahertz) radio frequency line of neutral hydrogen. Each advanced civilization must have discovered this line in the spectrum of cosmic radio radiation at an early stage of its development. Hydrogen is the most abundant element in the universe, and 1420 [megahertz] may be considered the fundamental frequency of nature.[5]

Researchers later proposed looking in the region of hydroxl, a molecule made of one atom of hydrogen and one of oxygen, which is not far from neutral hydrogen on the radio spectrum. According to Eugene Mallove:

Since hydrogen and hydroxyl are the products of the decomposition of water, the frequencies near those they emit have been romantically dubbed the "water hole." SETI researchers suggest that galactic species might mingle in this region via radio waves, just as animals congregate at water holes on Earth.[6]

The first formal effort to realize the vision of Cocconi and Morrison began in 1960. Dr. Frank Drake, then at the National Radio Astronomy Observatory in Green Bank, West Virginia, created a receiver to detect signals in the 1420 megahertz range. He aimed his receiver at Epsilon Eridani and Tau Ceti, both between ten and twelve light-years from Earth.[7]

Drake's project was called "Ozma" after the Queen of the Land of Oz. The listening continued for several months, involving about 150 total hours of observation time. Early in the project, Drake thought he had detected an extraterrestrial signal, but it was coming from an Earth-based source.[8]

Drake's experiment used the 27 meter radio telescope at Green Bank, and required relatively little additional equipment.[9] It was unsuccessful in detecting extraterrestrial signals, but it did move the debate over extraterrestrial intelligence out of the realm of speculation and into the domain of science.

Ozma was an historic step in the history of space exploration, ranking perhaps with the initial flight of Yuri Gagarin or the *Apollo 11* landing on the moon. Space travel, too, had been a topic of speculation for thousands of years, but people like Gagarin and Neil Armstrong made it a reality.

During the early days in developing this radically new field of space exploration, SETI scientists began to discuss the various ways in which an extraterrestrial signal might be detected. There appeared to be three obvious possibilities:

1. **The beacon option**
2. **The eavesdropping option**
3. **The broadcasting option**

The beacon option assumes that a civilization is sending out a signal with the intention that it be received by another civilization at or near our level of development. The signal might even be directed at specific planets. From a technical point of view, the beacon option is the most promising. If a signal is aimed directly at Earth, we can assume that the senders will take into account the obstacles preventing us from receiving it. Therefore, it is reasonable to think they would send out a signal that is easily received and recognizable as artificial.

The eavesdropping option is just what its names implies. Extraterrestrial civilizations might be communicating with one another and ignoring us, but we could still "tune in" and pick up their transmissions. Frank Drake suggested that we could analyze a characteristic signal repeating itself consistently against the backdrop of cosmic radio noise. Such a signal would most likely be of artificial origin, and we could "listen in" on it.[10]

The broadcasting option is the final possibility. Our own civilization has been transmitting radio and television signals into outer space for some sixty years now. Traveling at the speed of light, those original signals have now reached all stars within some sixty light-years of us. A civilization on a planet circling one of those stars might pick up a television program and realize that it emanated from another society at a given level of development.

The receiving civilization might send out a beacon in response to the signal from Earth. That is in fact the premise of Carl Sagan's fictional account of the first SETI success, a novel called *Contact*. In the book, the extraterrestrials receive one of the first television signals ever generated, a picture of Hitler presiding over the 1936 Olympic Games. They respond by sending the same picture back, as well as dramatic new information of their own.[11]

Detection Options

The modern SETI movement has not limited itself to searching for electromagnetic signals. The search for extraterrestrial intelligence is fundamentally an effort to find areas of the universe in which "matter is being transformed into mind." The manifestation of that process is the utilization of energy on a broad scale. Energy utilization includes communications, but is not limited to it.

Early SETI detectives realized that it might be possible to detect other unusual patterns of energy usage across interstellar distances. The Soviet scientist Nikolai Kardashev pointed out that increasing control of energy sources is the mark of an advancing civilization and that a culture's ability to communicate will be proportional to the energy it controls.

Kardashev hypothesized three scenarios for the development of what he called "super civilizations." His "Type I" civilization would be at a similar stage of development as modern terrestrial civilization, having gained control of most of the energy sources available on its own planet of origin. A "Type II" civilization would have reached a level at which it controlled the energy output of its own sun. A "Type III" civilization would have gained control of the energy output of its entire galaxy.[12]

How would a Type II or Type III civilization gain that much power over nearby energy sources, and what would they do with it? Physicist Freeman Dyson of Princeton University's Institute for Advanced Studies, has imagined the engineering feat that would bring about a Type II phase of development.

He wondered what terrestrial intelligence would do if it were to begin development of a Type II civilization. Dyson proposed that we might dismantle one of the larger planets in the solar system, such as Jupiter, and use it to construct a shell with a radius equal to the Earth's orbit and surrounding the sun, Mercury, Venus and Earth.[13]

That would bring Kardashev's Type II civilization into being: None of the sun's energy would be radiated into space, and every particle of it would be captured by the inner wall of the "Dyson Sphere." Dyson speculated that human beings would live on the inner wall, freeing them completely from the restrictions inherent in living on a planetary surface.[14]

Now, from the SETI perspective, it's not the inside of the sphere that matters, it's the outside. Dyson asked, "If such a sphere were constructed, how would it look from the outside?" According to him, it would be a large object equal in size to the Earth's orbit, with a high surface temperature, re-radiating the energy of the star inside it in the infrared spectrum. Since infrared radiation can be detected on Earth, a civilization utilizing a Dyson Sphere could be found as part of a search program.[15]

Since 1960, several searches have attempted to detect Type II or Type III civilizations, including some that specifically looked for Dyson Spheres.[16] In the mid-1960s, astronomers thought they might have found evidence of such civilizations. Two radio sources, CTA 21 and CTA 102, were sending out enormous amounts of energy in regions already identified as likely wavelengths for interstellar communications. After a period of observation, astronomers concluded that CTA 102 was a quasar, a natural source of intense radio energy. The nature of CTA 21 remains uncertain, but it is no longer considered a high priority for research.[17]

Type II and Type III civilizations might be detectable, even if they don't build Dyson Spheres. Natural systems "run down," or become "entropic," over time, moving toward what is called an equilibrium state. The presence of life and intelligence shows up as a disturbance, or disequilibrium in localized regions.

Viewing Earth from space, an observer would see evidence of entropy reduction caused by life in the high levels of oxygen in the atmosphere. Regions of increased entropy because of pollutants would also be observable in the same atmosphere. The evidence would be confined to a single planet, however, and would not be easily visible from another star system.

However, since Type II and Type III civilizations supposedly operate on a much larger scale, there ought to be more easily detected evidence of their activities. "Entropy pools"—areas of entropy reduction surrounded by areas of increased entropy—would appear as areas of disequilibrium that could not be explained as natural processes.[18]

Space-based telescopes will improve the search for extra-solar planets, and the ability to examine them. In 1983, the space-based IRAS (Infrared Astronomy Satellite) revealed that many stars had clouds of particles surrounding them, emitting infrared radiation—something astrophysicists would expect to see when looking at a planetary system.

IRAS was placed in Low Earth orbit, the site that is being planned for a number of "Great Observatories," or space-based telescopes, scheduled to be launched by the United States in the next decade. NASA has also proposed that one of its primary goals in the twenty-first century is the building of an observatory on the far side of the moon. That could be a boon to SETI, because it would provide a stable platform for optical telescopes beyond Earth's atmosphere and a listening post with minimal interference for radio telescopes.

From the time of Frank Drake's first Ozma experiment in 1960 until mid-1987, nearly fifty major SETI projects have been attempted around the world. In the first thirty years, many of the possible strategies discussed earlier were tried in one form or another. For example, a Russian effort at eavesdropping took place from 1972 to 1974, looking at the entire sky for 150 hours. Then, two scientists named Knowles and Sullivan attempted another eavesdropping project in 1978 in a five-hour observation of two stars.[19]

Beacon searches have included observations of seventy solar-type stars within forty-five light-years of Earth from 1972 to 1976, along with several all-sky surveys, such as that undertaken and continuing at Ohio State University's "Big Ear" site, which began in 1973.[20]

In 1975 and 1976, Frank Drake and Carl Sagan searched for evidence of Type II civilizations in what is called the Local Group of galaxies, and in 1978, three researchers looked at twenty-five globular clusters in search of Type II and Type III civilizations. In 1980, another scientist, working at a NASA site, observed twenty solar-type stars, trying to detect excess infrared radiation that might have emanated from Dyson Spheres.

There has even been a search for evidence of the dumping of nuclear wastes into a star's atmosphere. The idea is that advancing extraterrestrial civilizations will discover nuclear power, and that they will have problems with disposing of the resulting wastes. Francisco Valdes and Robert A. Freitas, Jr. also conducted a search for evidence of widescale nuclear fusion usage by extraterrestrial civilization.[21]

The work of Dr. Paul Horowitz at Harvard University remains one of the most important projects to date. Known as Project META, it is sponsored by the Planetary Society and received seed funding from Stephen Spielberg, the originator of *E.T.* META has been searching the sky visible from the Northern Hemisphere continuously since 1985 at a series of different frequencies. Recently, plans were announced to begin META II in the Southern Hemisphere as well.[22] META utilizes a multi-channel analyzer approach that is a precursor to that which the NASA project will employ, and Horowitz has worked closely with many of the NASA scientists in the past.

The projects brought on line during the modern SETI era illustrate a range of activities in terms of observers' nationalities, search strategies, and funding sources. Not all of the projects are "big science," with significant funds behind them, by any means. For example, Dr. David Latham of the Harvard-Smithsonian Center for Astrophysics informed me about Robert W. Stephens, who conducted a five-year search at Hay River in the Northwest Territories of Canada, using two large surplus satellite dishes.

The 1990s may well usher in a new era as governments, led by the United States, begin to take SETI much more seriously.

The NASA Project

"When the NASA SETI Microwave Observing Project starts, it will exceed all previous searches combined in the first half-hour of operation." That's NASA's prediction about the impact of its project, planned to begin in 1992.[23]

NASA, well known for its spectacular manned and unmanned missions to space, is now entering the SETI field in a big way. The agency's program has been an ongoing activity within the Life Sciences Division since 1982. As part of the Planetary Biology Program, its mission is to help understand the "origin, evolution, and distribution of life in the Universe."[24] NASA is being assisted on the project by the private, non-profit SETI Institute, an organization staffed by many SETI pioneers, including Frank Drake. The program has completed its five-year research and development period, and is moving toward operational status.

What makes the NASA project different from those that have gone before? It is primarily the opportunities afforded by new computer technology, rather than by radio telescope technology, that promises to transform the SETI field. To keep costs at a minimum, existing radio telescopes, such as the Deep Space Network facility in Goldstone, California, will be used for the search.

A new dedicated computer system known as the Multi Channel Spectrum Analyzer (MCSA) has been designed for NASA by Jay Duluk of Silicon Engines, and Stanford University. The MCSA can scan fifteen million channels simultaneously with various spectral resolutions. It is critical to the project because, like Project META, it is being designed to make many of the decisions about the value of what it picks up.[25] If a human operator had to make those decisions, the search would be restricted to a much smaller volume of space because of the time it would take to analyze all the data coming into the system.

The project points to the only possible future for SETI work, which will require extensive automation to be effective. According to NASA:

> The ultimate goal is to design automated SETI systems that require minimal operator assistance during routine observing, yet bring the operator on line the moment that the system cannot assign a signal to a known origin, human or natural.[26]

The program is divided into two parts:

The All-Sky Survey will search the entire "celestial sphere." It will span a wide frequency range from 1,000 to 10,000 megahertz, searching for beacons. This survey, which will take five to seven years, will cover ten thousand times more frequency space than all previous efforts, and will be about three hundred times more sensitive.[27]

The Targeted Survey will look for weaker signals emanating from some eight hundred to one thousand sun-like stars within about eighty light-years of Earth. This survey will focus on the 1,000 to 3,000 megahertz band, with special emphasis on stars within twenty light-years of Earth.[28] If these stars have planets and have evolved advanced civilizations, they may have detected our earliest television signals, and might be responding to us now.

The number of targets covered over a three to five year period will be much larger than previous, smaller searches and the range of frequencies thousands of times greater.[29] The difference between previous efforts and this project is based on a number of factors. According to a NASA description:

> The increased number of targets, the greater frequency range, and the wider variability of detectable signals make the SETI Microwave Observing Project 10 billion times more comprehensive than the sum of all previous searches.[30]

Dr. Kent Cullers of the NASA team focuses on the three components of the project that make this search different from those that have gone before:

> One [component] is that we will use the biggest antennas available on Earth to look at sun-like stars. . . . The second is that we will search the whole sky over a wide range of frequencies with slightly smaller antennas than the biggest possible, but [we] will do a whole sky search at higher sensitivities than have been done before. The last [component] is our computational capability. We will be able to process about one Encyclopedia Britannica full of random noise per second for five years at each one of our sites.[31]

As Cullers points out, processing an entire encyclopedia's worth of information every second is essential when you are looking through all that random noise for the one pattern that says, "Hi, we're the alien guys."[32]

Dr. Michael Klein is Deputy Project Manager for MOP, working out of the Jet Propulsion Laboratory (JPL) in Pasadena, California. He differentiates the NASA project from META in that MOP is looking at a very broad bandwidth. MOP does not try to guess in advance which frequencies will be used by extraterrestrials. META, by comparison, is a "preferred frequency" search, and is "the best of its type," in Klein's words.[33]

The NASA project also offers the ability to lock onto signals and examine them in "real-time," another major advance over many of the previous projects (other than META, which can also lock on). Earlier searches have detected what are known as "anomalous" signals, that couldn't immediately be explained as natural or known artificial signals. However, the earlier technologies required researchers to return to the same spot in the sky days or weeks later and look for the signal, analyzing the data after detection.

According to Kent Cullers and Peter Backus of the NASA project, the inability to work with real-time information made the data in some earlier projects less valuable than it might have been.

For example, Backus says:

> The researchers would record all the data on magnetic tapes and process it later on . . . and perhaps find an interesting signal six months after the observations. Then you would have to try to confirm that by re-applying for telescope time, going through a review process, getting assigned time. . . . You get back to the telescope, and now the equipment is different, and you have to re-configure it and try to duplicate the observations. When you attempt it, the signal is no longer there.[34]

Having all of these variables intervening in the process led to doubts about the results, says Cullers:

> That is no way to do SETI. Clearly, you just have to throw that data out. There's nothing you can do with it. You can't reconfirm it, you can't prove anything about it. Even if we found the real thing, no one is going to believe someone else's data.[35]

So far, none of the observing projects has been able to find an anomalous signal in the same spot in a succeeding observation period. The problem of "anomalous but non-repeating signals" is an intriguing frustration for SETI scientists.

The anomalies resemble the goal of the search in that they are strong spikes standing out from the background noise of the universe. However, their refusal to repeat themselves, ever, means that they do not meet one of the SETI community's fundamental criteria for detection.

The Ohio "Big Ear" site reported an anomaly that became famous as the "WOW" signal, because a researcher wrote the word "WOW" next to it on the printout where it appeared. However, the signal did not appear during the next observing period, nor has it appeared again, so it is considered an unsolved mystery.[36]

Michael Klein, in discussing this incident, emphasizes the importance of handling potential false alarms with care. Klein says radio astronomers do find "little things that go bump in the night," which can't be explained right away. He asserts that "Our policy is that if you can't confirm it, it doesn't exist."[37] These are strong words, but they focus on the deep concern that SETI scientists feel about the issue.

After eighteen months of operation for Project META, Paul Horowitz wrote a report for the Planetary Society's magazine, in which he said that there had been three anomalous signals recorded during that time, none of which could be confirmed. Echoing Klein's comments, Horowitz observed that, "Irreproducible science is not science at all; thus we are obliged to discard these events, without a good explanation of their cause."[38]

After several years of operation, Horowitz has come to believe that the anomalous signals are almost certainly artificial, but that they are coming from Earth-based sources.[39]

Credibility is established in other ways, including peer review of methodology. Peer review is a standard procedure within the scientific community, used to judge whether a project ought to be undertaken. MOP has been through many evaluations and has been approved in each case.[40]

Not everyone in the scientific community agrees that putting a substantial level of resources into one government project is the right way to do SETI, and there are continuing disagreements on overall strategy. However, the NASA project appears to have enough support to give it credibility if it reports a successful detection or contact. That is an extremely important factor in terms of impact, because it means that the scientific community is likely to take seriously any results announced by the NASA team.

Field tests have also been performed on both strong and weak signals to test the capability of the receiving equipment and the computer equipment. To test the technology's ability to capture strong signals, the radiotelescope located in Goldstone, California was aimed at the *Voyager 2* spacecraft, which was near the planet Uranus at the time. The test equipment array included a prototype of the 74,000-channel analyzer (MCSA), and the test was successful.

To test the ability to pick up weaker signals, the same equipment was used to search for the signal of *Pioneer 10*, which was then beyond the orbit of Neptune, some 3.8 billion miles away from Earth, and that test was also successful.[41]

Pictures of a spectrogram from the *Pioneer 10* experiment show us what an intelligent signal from another solar system would look like, with tell-tale spikes jutting above the wiggly lines of background radio noise. The *Pioneer* test is significant because the transmitter on the spacecraft emits one watt of power, about half the power of a miniature Christmas tree light![42]

Another factor affects the success of the NASA project (and other SETI projects), but it is linked with human behavior, rather than with science and technology. MOP is budgeted at about $105 million over a ten-year period, which is low compared to many other human activities, but high enough at a time of budget deficits and mounting social problems.

The NASA budget is always at risk from those who think the money would be better spent on programs closer to home, and no one can guarantee that any program will be funded throughout its lifetime, including this one.

The ultimate issue is not the SETI funding, but the commitment of the United States and the people of planet Earth to the exploration of the universe. If we are serious about SETI, about discovering who we are in relation to our universe, we will "put our money where our mouth is" and see the project through to its logical end.

What If?

The NASA project is important for its scientific and technological approach to SETI, and it is also breaking new ground in its approach to the social and political issues surrounding the search.

The SETI Office intends to begin an outreach effort to the public before the project goes on-line in 1992. They plan to involve schools, for example, in subsidiary projects, allowing students to feel that they are participating in the project.[43]

Bob Arnold, formerly with the Voice of America, has joined the NASA team to work on the outreach program. He sees SETI as a powerful way of teaching young people about the sciences from an interdisciplinary perspective. According to Arnold, "The topic of SETI is exciting, with excellent potential for motivating people of all ages and backgrounds to learn the scientific concepts it draws upon."[44]

Many different segments of society could become involved in a pre-detection planning process that would stimulate them to think globally while relating contact to their own local context. In this way, the SETI Factor, long an unconscious influence on human thought, would begin to move into our social awareness, becoming a conscious force for planetary unity and human evolution.

To many of us, the most intriguing question is, "What do we do if it works?" We must think about how we would respond, on a global basis, to acquisition of a signal. In an effort to provide an answer, several people have begun work on a "Post-Detection Protocol."

The NASA project has developed its own draft protocol, which may be followed if a signal is detected by MOP. However, in 1987, if adopted it would control the actions only of researchers working on that project. In the meantime, an international effort, led by Michael Michaud of the U.S. State Department, has begun. It is aimed at obtaining the support of all SETI researchers around the world for a global post-detection protocol. Here, we will focus on the NASA draft protocol as a specific example, but it is generally consistent with the proposed international agreement.

Post-Detection Protocols

Looking at the post-detection protocols opens the door to the emerging new field of "interstellar relations." SETI researchers are not in agreement on everything that should be done in this area, but there is an emerging consensus on one point, which is that any extraterrestrial signal belongs to the world and should not be suppressed by those who acquire it.

There is concern within the SETI community that those in the government responsible for national security will try to classify the fact that a signal has been acquired, as well as its content. The NASA protocol takes a different approach, declaring that:

> The detection of an extraterrestrial civilization is a discovery with such profound implications that it transcends national boundaries and should be the property of all mankind.[45]

At first, the statement seems almost too obvious. Contact with extraterrestrials is clearly a matter of global concern, and it is hard to imagine that they would intend the message for one nation on a planet they can't even see. On the other hand, the project is funded by American taxpayers' money, not by the people of the world. Suppose that the signal contains information of tremendous value? Is the United States obliged to share that information openly and without any filtering or editing?

The use of government money can also be the foundation of an argument that results ought to be shared. The American space program has been relatively open, and sharing the SETI results fits with that tradition. Should the same principles, however, apply to private SETI projects, such as that of the Planetary Society? Dr. Louis Friedman, Executive Director of the Society, expresses reservations about the protocol:

> We are watching it with interest, but we aren't sure it's as vital as doing the work. We would like to see SETI be considered as a scientific experiment and handled by the scientific method. There are established ways of conducting scientific experiments and reporting on them. . . .[46]

As private sources fund more of the space exploration field, we have to ask what happens if an entrepreneur or corporation invests millions of dollars in SETI specifically to market the information generated by a search. In that case, it would be contradictory to talk about giving away the results. All questions concerned with the exploration of outer space, such as "Who owns the moon?" put strains on Earth-bound legal concepts, and SETI is no different.

A second emerging principle is that SETI researchers must strive to maintain their credibility by avoiding premature announcements of success. This principle provides checks and balances against the first principle, which assumes rapid dissemination of results.

The NASA project builds in several levels of verification to prevent false alarms. At the first level, automatic verification procedures are designed into the computer system software to eliminate radio frequency interference (RFI), equipment malfunction, or distant Earth-launched spacecraft as sources.[47]

In addition, the ability to lock on to an anomalous signal and start analyzing it in "real-time" will go far toward preventing false alarms. The equipment will hold the signal until the rotation of the earth causes the signal to appear to "set," like the sun. This capability will provide the SETI teams with a much more sophisticated understanding of each signal than has been possible in the past.

The next levels of verification require that the discovery team inform the SETI Office, which in turn informs other MOP sites of the find. Other observatories are informed so that they can attempt to verify the signal's presence as well.[48]

After the signal has "set," the discovery team is required to reload new system software for use in the next observation period, insuring that the software itself is not announcing a false detection. If the signal is recaptured in the next observing period, the SETI project office informs NASA Headquarters in Washington. The NASA Administrator notifies administration officials and congressional offices as appropriate. NASA Headquarters then prepares a news release, emphasizing that a signal has been acquired, but that it has not been confirmed to be "ETI."[49]

Even though the entire project is highly automated, humans make the final determination that the signal is a sign of extraterrestrial intelligence—in the end, it's a judgment call. The plan is for a group of experts (scientists, engineers, and others familiar with the goals of the project) assembled by the project team to determine that the search has been successful. These experts would meet and review all available data and they would then come to one of three possible conclusions:

1. **The signal is clearly astronomical (natural) in origin**
2. **Further tests are required to determine the signal's origin**
3. **The signal is clearly being generated by an extraterrestrial intelligence**[50]

If the panel decides that the signal is astronomical in origin, an International Astronomical Union (IAU) telegram is sent out to announce the discovery, a standard operating procedure for major finds. NASA Headquarters would also send out a news release announcing the discovery, and if further tests are required, NASA sends out a news release, emphasizing that the signal is most likely of astronomical origin.

If the panel concludes that the signal is definitely a product of extraterrestrial intelligence, the procedure becomes far more complex. First, the NASA Administrator would be notified of the decision, and he or she informs the Executive Branch and Congress. NASA Headquarters then prepares an announcement and sets up press conferences to announce to planet Earth that "We are not alone."

The draft protocol assumes that the President, NASA Administrator, or both, will make a formal announcement at this time. In addition, the scientific and technical results would be published as soon as possible in the appropriate journals.

In the meantime, the project team continues to monitor the acquired signal, and all of the resulting data will be pooled. The NASA draft protocol has recommended that analysis and interpretation of the signal content be done by an international team of scientists, an approach that further insures the information will become the property of all humankind.[51]

The final step in the protocol calls for re-examination and review of the entire project to determine whether any new actions should be taken. The project is aimed at detecting only one signal, but once a signal is detected, the thinking must diverge. The team will want to allocate sufficient resources to the analysis of that one signal, and they will begin to wonder, "How many more are there?" As former project manager Lynn Harper says, "If there is one signal, there may be many."[52]

For example, the targeted search is examining some eight hundred nearby stars, and the sky survey is looking at the entire sky. Suppose that a signal is acquired as the equipment focuses on the one hundredth star in the targeted survey. Obviously, the project team would want to sit down and ask itself, "What shall we do next?" It would be impossible simply to continue the project plan as if nothing has happened.

The sudden burst of publicity and public interest that follows the revelation of success will also force any successful SETI researchers to re-think their planning. Overnight, SETI would go from being a relatively obscure and esoteric activity to a matter of the utmost

interest to everyone on planet Earth. The researchers will instantly
acquire not only a signal but also a coterie of advisors, telling them
what to do next. (In 1986, the shuttle program was generating so
little interest that the launch of the *Challenger* was not carried on
network television. However, when it exploded, NASA and the
program got far more publicity than either had received in many
years.)

If a valid signal is acquired, humanity will move from the "pre-
contact" to the "post-contact" era, and the issue of impact becomes
an immediate reality. NASA has expressed interest in this matter,
but the agency cannot take full responsibility for the outcome of its
work. That is a matter for the people of Earth to consider.

Consideration ought to begin soon, however. There are already
several SETI projects underway. Project META is expanding, and the
NASA project begins in two years. We must now begin to confront
the possibility of contact with extraterrestrial intelligence within
our lifetimes. Once it happens, we will be playing "catch-up ball"
on an issue that demands the maximum amount of forethought.

After They've Said Hello

Conspicuously absent from the post-detection protocols is a step
that includes responding to an acquired signal. The reason is directly tied to the level of social evolution that our planet has reached
in the late-twentieth century. To put it quite simply, the question is,
"Who speaks for Earth?" And the answer is, "Everyone and no one."

We have grown accustomed to news reports that read, "The United
States said today . . ." or "The Kremlin announced that . . ." Nations
"speak" for themselves, but there is no tradition making it possible
to say "The Earth announced today that it had made its first contact
with extraterrestrial beings."

Dr. Jill Tarter, Project Scientist for the NASA project, has taken
primary responsibility for developing the NASA draft protocol, and
Michael Michaud of the State Department has been a leader in
pushing for an international protocol. The international agreement
contains the same basic principles of rigorous confirmation and
widespread dissemination as the NASA protocol requires, but is
adapted to serving a larger research community.

Tarter and Michaud have found that consensus on handling detection is difficult, but seems within reach. However, there is no consensus on response.[53] The International Institute of Space Law is currently considering the issue of how to respond, if at all. For now, the "Declaration of Principles Concerning Activities Following the Detection of Extraterrestrial Intelligence" requires that no response be made without international consultation.[54]

No one knows how to respond, partially because no one speaks for Earth as yet, but also out of a conservatism based on fear of the unknown. In explaining why no one on Earth is sending out beacons, Dr. Bernard Oliver says:

> There are people who are fearful that if we announce our presence, they will descend upon us. So their philosophy is to 'keep quiet in the jungle or get eaten.' You have to convince those people or enough of the populace that it is a good idea before you go ahead and do it in a democratic society. On the other hand, listening has no such hazards in anybody's mind.[55]

The discussion of post-detection protocols is a positive development in the ability of human beings to anticipate and manage the impact of technology-related social changes. It also represents a change in thinking that has transpired in the past two decades.

According to Charles Redmond, public affairs officer for NASA: "As a group, one of the things we've been trying to do is figure out, 'Are there questions we haven't asked in the social arena that we now have the sophistication as a technical society to ask?' "[56]

As an example, Redmond says that NASA gave relatively little thought to the impact that returning to Earth would have on the *Apollo* astronauts. Today, however, NASA realizes that its technical achievements have social impact, and these must now be taken into consideration.[57]

SETI is likely to exert a continuing and growing influence on the public mind over time. Thinking about contact with extraterrestrials forces people to consider the matter globally, but it does not guarantee consensus. There must be vigorous and widespread debate before the world can decide how to deal with extraterrestrial intelligence.

However, that debate itself can forge a new level of unity on the planet, as humans work to resolve this intensely complicated issue. Contact could come any day now, and it is not too early to begin "getting ready for SETI." Eventually, a response may be necessary, perhaps the most important message in our planet's long, raucous history.

What would you say?

Chapter Six

The Probabilities of Contact

> *Technology has provided the first opportunity to explore*
> *our vast Milky Way galaxy. We can now study our*
> *universe to ascertain how frequently the process of*
> *Cosmic Evolution has led to the development of other*
> *technologies. In this way we can begin to attempt to*
> *answer the question "Are we alone?"*

> —"Cosmic Evolution," NASA SETI Project

SHOULD WE BE DISAPPOINTED BY THE APPARENT LACK OF SUCCESS during the first thirty years of the modern SETI movement? Opinions vary widely. Some believe that the lack of a signal means that nobody's there. Others think any conclusions are premature because we have searched so little of the universe.

When science is done well, nothing is lost. Experiments that fail to produce desired results still educate the experimenters. SETI is good science, and it has increased human knowledge of the universe enormously. The big prizes, discovery of life or contact with intelligence, have proven elusive, and that has to be a disappointment. Nevertheless, several conclusions can be drawn from the searches of the past three decades. These insights are important enough to discuss in some detail before moving on to analyze the probabilities of contact in the future.

First, we can conclude, in the words of Paul Horowitz, that our galaxy is not "teeming with (intelligent) life and radio signals."[1] According to him, the searches conducted even before Project META went on-line in 1985 were sensitive enough to confirm that conclusion.

Second, there are no Type II civilizations in the galaxy using enormous amounts of energy to communicate with Earth or with others. As Kent Cullers puts it:

> For all we knew in 1959, there were civilizations transmitting the power of an entire sun in the radio range, and we would never have known they were there. Today, we know that within the galaxy, there is nobody doing that, because our searches have been that sensitive.[2]

Third, the searches have detected something out of the ordinary many times, something tantalizingly close to the object of the search. The "WOW" signal, the "little things that go bump in the night," the "bold spikes" that appear once, and then never return—these anomalous signals may be something to be examined more closely.

The anomalies are not an isolated phenomenon, having been seen frequently at "Big Ear" in Ohio, and relatively frequently by Project META. However, SETI experts tend to dismiss them. Paul Horowitz is convinced that they are Earth-based signals, and Philip Morrison feels that they are simply not interesting enough to be seriously pursued.[3] However, the anomalous signals may be an important discovery made during the first thirty years of SETI, and they deserve further investigation.

The probes and manned explorations of the solar system have produced many surprises, such as the discovery that water was once abundant on Mars, and that there is volcanic activity on the moons of the outer planets. The life detection experiments on Mars remain intriguing, while the discovery that organics are relatively abundant in the solar system is downright exciting.

Nevertheless, the solar system simply isn't crawling with life as we know it or extraterrestrial intelligence like ourselves. This discovery is similar to finding that there are no Type II civilizations beaming messages to everyone in the Milky Way galaxy. It doesn't mean that life and intelligence aren't present, but merely that neither is obvious today.

The exploration of the solar system has triggered a major shift in thinking about the presence of life and intelligence close to Earth. Until the mid-nineteenth century, many believed that some or all of the planets of the solar system, including the sun, were inhabited.

For example, Johann Elert Bode (1747–1826) believed that extraterrestrials might live inside the cool core of the sun, looking out into the universe through sunspots:

> Who would doubt their existence? The most wise author of the world assigns an insect lodging on a grain of sand and will certainly not permit . . . the great ball of the sun to be empty of creatures. . . .[4]

Bode's thinking anticipates the Dyson Sphere concept by over one hundred years, but he is undoubtedly wrong. A pattern persisting in this field over time—speculation retreating in the face of evidence based on observation—has repeated itself regarding the issue of intelligent life within the solar system. Those who clung to the view that planets other than Earth were inhabited made their last stand in the great Mars controversy at the end of the nineteenth and beginning of the twentieth centuries.

SETI researchers now assume that there are no extraterrestrial civilizations nearby because conditions do not seem to warrant it, and we would have heard from them if they were there.

Saturn's moon, Titan, will continue to be an object of interest for astronomers. Titan's relatively thick atmosphere appears to harbor organic compounds, and computer models of the primitive Earth and present-day Titan agree in many respects. Titan has not yet evolved from organic compounds to cells because the temperature of minus 290 degrees Fahrenheit keeps all of the water on the moon locked up as ice. To take the next step in evolution, the ice would have to melt and become available to link up with carbon molecules.[5]

Mars also will continue to hold a unique place in our thoughts about extraterrestrial life and intelligence. The planet has fascinated the popular mind for many years, and it is intriguing to planetary scientists as well. Mars is half the size of Earth, but it has about the same total land mass, since it lacks oceans. Mars also supports a thin atmosphere and polar ice caps that advance and recede with the seasons.[6]

In the late-nineteenth century, an Italian astronomer named Schiaparelli observed interlacing lines on the surface of Mars. He called them *canali*, which translates into English as "grooves." However, the term was often mis-translated as "canals." Percival

Lowell, an American astronomer, believed that the lines formed a network of artificially-constructed canals. He hypothesized that they had been built by an ancient, dying Martian civilization in order to bring water from the polar ice caps to the arid regions of the planet.

An intense debate ensued over whether the "canals" even existed, or were an optical illusion, and an even fiercer controversy erupted over Lowell's claims. He mixed fact and fancy into an unscientific brew that eventually became embarrassing to the astronomical community.[7]

Lowell's ideas have been discredited, but fascination with the planet and its mysteries remains. Carl Sagan, writing about the controversy, suggests that the canals must have been hallucinatory in nature, because nothing can be seen today that really corresponds with them.

Sagan adds:

> The canals of Mars seem to be some malfunction, under different seeing conditions, of the human eye/hand/brain combination. . . . But this is hardly a comprehensive explanation, and I have the nagging suspicion that some essential feature of the Martian canal problem remains unresolved.[8]

Humans will go to Mars for many reasons, and they will discover many new things there, perhaps even evidence of ancient life and civilizations. The *Magellan* probe to Venus, and the *Galileo* probe to Jupiter, both launched in 1989, will generate volumes of new information about those planets.

Humans will go to Titan, Europa, Io and every other part of the solar system, and each journey will provide new data about how life and intelligence develop. We may even create life on Mars, Venus, or Titan by bringing terrestrial life forms to those planets and intervening in the evolutionary processes there.

Based on what is known now, it seems slightly possible that primitive life may be found somewhere in the solar system, but highly improbable that advanced intelligent civilizations will be.

While life and intelligence may be scarce in our solar system, the search for extra-solar planets indicates that planets may be abundant in the universe, possibly providing millions of potential life-sites.

The detection effort has, then, produced much new information, and human thinking about the presence of extraterrestrial intelligence has shifted dramatically in the past thirty years. Earlier researchers confidently, but erroneously, predicted abundant life and intelligence in the solar system. Today's advocates now hope to find it among the stars. The next phase in the detection process is to determine whether that expectation will also be shattered or confirmed.

As that phase is approached, the tantalizing questions remain:

1. How likely is contact with an extraterrestrial civilization?
2. What is the likely impact of contact on us?

These questions must be addressed from an interdisciplinary perspective because they are concerned with the structure of the universe, the origin of life, evolutionary processes, and the nature of intelligence. Issues concerning the probability of contact fall primarily in the realm of the physical sciences, and require knowledge of many fields in order to think intelligently about the question.

The impact of the search, and especially of contact, moves into the domain of the social sciences and humanities, raising questions about the origins of society and civilization, interactions of societies at different levels of development, dissemination of information, belief systems and perceptions of reality.

The remainder of this chapter deals primarily with the probability of contact, while the next chapter will cover the impact of the search, and of contact in particular.

True probabilities can only be estimated when a great deal is known about the system being studied, which is not the case with SETI. For example, the probability that certain people will become health risks can be calculated by an insurance company because historical information is available about types of people and the diseases they tend to contract.

The people working on SETI don't like to talk about probabilities of successful contact because there are simply too many unknowns.

Dr. Bernard Oliver believes that the probability of success depends largely on the types of signals that might be generated:

If another civilization is beaming something at us, we will pick it up. I doubt that we will have enough sensitivity to eavesdrop. So that lowers the probability of success in my mind considerably, because I think that if we are not sending out beacons, why should they?[9]

Dr. John Billingham agrees that the probabilities are difficult to calculate, but he believes that as time goes on, the likelihood of success increases:

If I had to guess on statistical grounds, I would say that during the early years, the probability is not high. As we go on, we will cover more and more of the available "search space," and during the whole ten years, I like to feel there is a reasonable probability of success.[10]

Dr. Billingham's statement points to the importance of distance and time in considering the likelihood of contact. Let's examine them more closely.

Distance

The likelihood of contact with intelligent life diminishes as we move outward from the earth to the edge of the solar system, increases as the search moves into the Milky Way galaxy to a distance of a hundred light-years or so, and then decreases again. There are a number of candidate sun-like stars within the search space selected by NASA's targeted search, and the available equipment will be powerful enough to examine each of them carefully over a reasonable period of time.

Beyond that distance, but still within the Milky Way galaxy, the likelihood of detection and/or contact diminishes. Lowering the estimates beyond about one hundred light-years is not based on assumptions about the presence or absence of life and intelligence at those distances. Rather, it is based on limitations in the resources available for searching, and assumptions about the extent to which we have made ourselves known.

Fairly weak signals from star systems within a range of fifty to one hundred light-years can be acquired by today's listening equipment. The farther one goes beyond that distance, the stronger the signal sources must be before they can be detected. Moreover, the searches of the past thirty years have already shown that there are no super-strong signal sources emanating from all corners of the galaxy.

The Milky Way galaxy itself is huge, spanning a diameter of some 100,000 light-years. The galaxy nearest to the Milky Way is 170,000 light-years away. At these distances, detection becomes increasingly difficult, though certainly not impossible.

Time

The likelihood of detection and/or contact increases with time. If human beings continue to explore the universe, humanity will expand outward, diminishing the distances between ourselves and potential sites of life and intelligence.

If life does exist somewhere in the solar system, it is likely to be found in the next few hundred years, as humans explore and settle the solar neighborhood. If intelligence has been here, or if it arrives, the possibilities for detection and contact also increase as the years go by, and we become a multi-planet species.

The sophistication of SETI in all its forms will also grow. Plans are already being laid for probes to go to the stars, space-based telescopes to detect extra-solar planets, and observatories on the moon. Computer technology shows no signs of becoming less so-phisticated, and signal processing is likely to produce better and better "electronic ears" for human listeners.

Critics of SETI say that interstellar migration/exploration would have brought extraterrestrials to our solar system by now. If they exist, and something unknown is slowing them down in their expansion, then the likelihood of their arrival increases with time, as does the likelihood that evolution will produce new spacefaring civilizations.

The reverse of that argument is that humans will be able to explore most of the galaxy within the same time frames quoted for extraterrestrials. The estimate of 300 million years needed for galactic exploration, quoted by Frank Tipler, is awesome to beings with life spans shorter than a century. However, it is a short time in cosmic terms, and it is the cosmic clock that counts.

The Drake Equation

The classic approach to thinking about probabilities of contact is to use the "Drake Equation," proposed at an early meeting of SETI pioneers by Dr. Frank Drake, originator of Project Ozma. The equation estimates the number of technological communicating civilizations that might exist within the Milky Way galaxy (the approach can be extended, of course, to the universe as a whole).

The Drake Equation only indirectly addresses probability, because the answer it generates—the number of technological communicating civilizations in existence—does not speak to the likelihood that contact will actually occur.

There are two widely used versions of this equation:

(1) $N = R_s \times F_p \times N_e \times F_l \times F_i \times F_c \times L$ (the original version)
or

(2) $N = N_s \times F_p \times N_e \times F_l \times F_i \times F_c \times L$

The equation looks intimidating, but once the eight key symbols are understood, it's a matter of common sense. As many SETI researchers have noted, it is not an equation with a clear answer, because it contains far too many unknowns. It should be seen perhaps as a model, a simplified way of thinking about the most important SETI questions. Here is how the equation looks when it is written out:

Number of Extraterrestrial Civilizations Able and Willing to Communicate = *(Number of stars in galaxy, N_s, or rate of star formation, R_s)* × *(Fraction of stars with planets)* × *(Number of planets ecologically suitable for life)*

×	*(Fraction of planets suitable for life where life has evolved)*	×	*(Fraction of "life- starts" where intelligence has appeared)*	×	*(Fraction of intelligent species who have created technical civilizations)*	×	*(Fraction of technical civilization: still in existence, F_L, or estimated lifetime of such civilizations, L)*

Let's now look at each of these quantities in turn. In this first discussion, we will use no numbers, focusing only on understanding the model. Then, to see how it works, we will illustrate how a relatively conservative set of assumptions can produce a fairly large number of civilizations. We will then show how different numbers will produce only one civilization, our own.

The equation's variables are stated in slightly different ways by various interpreters of it. For example, version (1) uses R_s, the mean rate of star formation averaged over the lifetime of the galaxy or the rate of formation of stars suitable for development of intelligent life, instead of N_s, the number of stars in the galaxy. It also uses L, the mean lifetime of technical civilizations, instead of F_L, the fraction of a planet's life during which a technical civilization exists.

If the same underlying assumptions are present, each version should yield the same answer. In our examples, we will use version (2) because it is somewhat easier to understand. It may sometimes be different from the original formulation by Dr. Drake, but it will remain consistent with the spirit of the original idea.

Let's look, now, at each of the quantities in the equation.

N: N is the answer generated by the equation; it stands, in Sagan and Shklovskii's words, for, "the number of extant advanced technical civilizations possessing both the interest and the capability for interstellar communications."[11]

This answer to the equation is straightforward, but a few points are worth noting about the definition. First, it refers to "extant," or existing civilizations, specifying that the answer will not include all those civilizations that might have reached the communicating stage, and then stopped communicating or became extinct. In addition, it refers to those civilizations with both the interest and capability to communicate. The definition implies that these civilizations are in fact making efforts to communicate.

N_s: N_s is the first number to be inserted into the equation on the right-hand side of the equal sign.

There are several different ways of thinking about this quantity. In more technical presentations, it is discussed as the mean rate of star formation averaged over the lifetime of the galaxy, and that is the most accurate description of it. A simpler way of stating it is to call it the current number of stars in the Milky Way galaxy.

This component of the equation is critical, for obvious reasons. Scientists assume that intelligent life initially evolves on planets, and planets are believed to exist only as companions to stars. For this reason, the number of stars in the galaxy must be the starting point for estimating the number of extraterrestrial civilizations in existence. The greater the number of stars, the greater the potential sites for life and extraterrestrial civilizations.[12]

F_p: F_p is the fraction of stars in the galaxy accompanied by planets. Whereas there is a good estimate for the number of stars in the galaxy, the estimate of those with planets is less certain.

According to Dr. David Latham of Harvard-Smithsonian Center for Astrophysics, it is strongly believed by many in the field that planets are ubiquitous throughout the galaxy.[13] Nevertheless, the confirmed existence of planets circling other stars remains unproven. Since such planets cannot be seen directly by optical means, they must be detected by indirect methods.

Dr. Bruce Campbell and his colleagues, working at the University of Victoria in Canada, have been developing this technique by studying a sample of eighteen nearby stars. They are examining fluctuations in the radial velocities of these stars as a means of determining the presence of planets.

How does this type of measurement reveal the presence or absence of planets? As planets orbit a sun, their gravitational attraction cause it to oscillate around a point called the "center of mass" of the solar system. The planets cannot be seen, but the motion of the star, responding to the planets, can.[14] Extraterrestrial observers could use these same techniques to determine if our own sun were accompanied by planets; perhaps someone is.

So far, the work of Campbell and his co-workers is encouraging to SETI advocates. Nine of the eighteen stars, or fifty percent, show distortions in their radial velocities that may be caused by planetary companions.[15]

Dr. Campbell believes that within the next several years, enough observations will have been conducted to announce whether planets are orbiting these stars. Then, as this line of inquiry continues, it will be natural to merge it with other SETI searches, because the stars with planets will be the best ones to target.

Another research team plans to mount a special telescope on space station *Freedom*, scheduled for completion in the mid-1990s, to conduct similar studies. Thus, it appears that by the year 2000, much more will be known about this component of the Drake Equation.[16]

N_e: N_e stands for the number of planets in a planetary system that are ecologically suitable for life to arise, primarily because they are properly situated within their own solar system. When we consider this part of the model, the analysis diverges from anything that is understood about objects outside our own solar system. Without knowing how many planets exist, it is impossible to know how many are suitable for life and/or intelligence.

If other solar systems are like ours, then the number of potential life-supporting objects in the galaxy might be high, because there are over thirty planets and moons in the solar system. Moreover, at least one planet is capable of supporting life and intelligence (Earth), and other bodies in the solar system appear to have some of the prerequisites for life as well.

Scientists have suggested that in any solar system, there may be a very narrow "zone of habitability" in which life not only can appear, but also flourish and sustain itself.

F_l: F_l stands for the fraction of planets suitable for life that have actually brought forth life. It asks the question, "Of those planets that can support life on a sustained basis, how many actually do so?"

The zone of habitability must be close enough to the sun to keep the planet warm, like the Earth, but not so close that a runaway "greenhouse effect" is triggered, as on Venus. The potential for life that exists on Titan may never realize itself because it is so cold there.

Once life gains a foothold, it appears to develop rapidly, and move from simpler to more complex forms. However, we still know of only one case where this has happened in our own solar system.

There may be many factors that affect the successful development of life. For example, Mars appears to be on the edge of the habitability zone, but its primary problem is its small size, rather than its location. Being half the size of Earth makes it harder for Mars to retain an atmosphere, which is probably why the Martian atmosphere is so thin.

Thus, whereas Titan has many of the necessary components for life, it cannot get living systems going, and Mars was unable to keep them going.[17]

F_i: F_i stands for the fraction of life-supporting planets where intelligence appears. It asks, "Of all those planets with living systems evolving on them, how many will go on to develop intelligence as well?" This question is quite difficult to answer, and even those scientists willing to grant that the galaxy and universe may be full of life are not so sure that intelligence directly follows from that.

In fact, this quantity in the equation has become something of a demarcation point for those who see a middle ground between the Assumption of Mediocrity and more pessimistic perspectives.

As Kelly Beatty, senior editor of *Sky and Telescope* magazine puts it:

> I think that . . . advanced intelligent life . . . is not unique to Earth, but is rare in the galaxy. The reason is that the complexities of biology are in some ways greater than the complexities of solar system formation. Life forms come up against many natural barriers to further evolution, and we may be rare in having come this far.[18]

According to Beatty, evolution may occur when species are under some kind of environmental pressure. If a species reaches a stable evolutionary stage, and is not further disturbed, it may remain within its ecological niche for a long time. The earth, however, has been a dynamic place for evolution to occur, for a number of reasons, according to Beatty:

> Our solar system happens to be one where there is a lot of stuff left over from the creation of the planets—comets and asteroids that continually bombard the Earth and re-shuffle the deck of environmental conditions.[19]

On a cosmic timescale, life appeared astonishingly early on the earth, then existed for a long time without developing a human level of intelligence. Mars developed many of the necessary elements for the evolution of life and intelligence, but the possibilities were not realized. Going from the potential for life to life itself and then on to intelligence is apparently not inevitable.

Thus, another solar system could exist with two or more planets primed for the evolution of life and human-type intelligence, and none of them would ever bring forth a species anything like ourselves.

Dale Russell, an anthropologist, has conducted instructive work in the field of *encephalization*. The term refers to the growth of the size of an organism's brain in relation to its body size. According to Russell, the larger the brain relative to the body, the more intelligent the organism will be, because more of the brain is freed to think rather than manage the body.[20]

Russell has traced the rate at which brain growth has occurred on Earth. He suggests that environmental factors can be determined that must be present in order for those rates to reach the level currently enjoyed by humans. He points out that if there had not been a sharp and unexplained shift in that rate 230 million years ago, human intelligence would not now exist even on Earth.[21]

Evolutionary theory looks at what is "adaptive," when new organisms or behaviors appear, and we would expect intelligence to appear if it helped an organism survive within a given environment. With hindsight, the advantages of intelligence in human beings are obvious, but it is not clear that intelligence as we know it will always be necessary on every planet that brings forth life.

Some experts have noted that intelligence may not always be adaptive. If the pun can be forgiven, there is a lot of "overhead" involved in maintaining a big brain. Also, there seems to be a correlation between encephalization and the period of dependency for the young of a species, putting a major strain on the adults and creating danger for the young.

There are organisms, such as sharks, that have functioned successfully on Earth for millennia while changing very little, and without what we would call intelligence. Big brains may be necessary on Earth for a special reason that would not obtain elsewhere. The earth seems to have periodic extinctions every 26 million years that wipe out large numbers of species, such as the dinosaurs. That may severly disturb the planet's evolutionary equilibrium and open the way for rapid encephalization, and it may be unique to Earth.

No one knows exactly why these periodic extinctions occur, nor whether the pattern would persist on other planets.[21] Other planets may reach an equilibrium point where brains do not enlarge and intelligence does not increase.

In Carl Sagan's words:

> Some think that the equivalent of the step from the emergence of trilobites to the domestication of fire goes like a shot in all planetary systems; others think that, even given ten to fifteen billion years, the evolution of technical civilization is unlikely.[22]

F_c: F_c stands for the fraction of planets with intelligent life where technological civilizations have appeared. Of those planets where intelligent life has emerged, how many have produced a technical civilization able and willing to communicate with the rest of the galaxy? There are grounds for both optimism and pessimism regarding this component.

Twentieth-century human beings living in industrialized countries take science and technology for granted. However, there is no guarantee that if a planet produces human-like life, it will also produce complex technical civilizations.

Consider tribal cultures on the earth, for example. The native

Americans of North America and the aboriginal peoples of Australia developed complex social structures, highly evolved cosmologies and unique philosophies of life. However, they showed no great inclination toward technological evolution, and it is not clear that they ever would have become a spacefaring civilization. In fact, those cultures put a high premium on maintaining balance and harmony with nature, which retards the kind of innovation and technological development characterizing modern society.

Consider further the example of the Inca of Peru, and the Aztecs of Mexico. By the fifteenth century, they had developed civilizations with major cities, armies, powerful religions and artistic accomplishments. However, they did not invent the wheel and other technologies that are now commonplace. Had they not been overcome by the Spaniards, would they have evolved further, and would they have become involved in a SETI-like process?

Cultures and civilizations, like species, can remain at certain levels of development or in given niches for millennia if they are not disturbed by outside forces. Terrestrial civilization was at an equilibrium point until recently and was transformed into a technology-driven society primarily by the Industrial Revolution.

That revolution emerged out of a complex pattern of changes in human thought, including the Renaissance and the Copernican Revolution, which brought a new cosmological system to bear, shattering the hold of the medieval Church on thought and action. Traditional cultures and organized religion resist modernization even today, and the revolution is not complete worldwide.

It is not only life that must appear on a planet, nor even intelligence and civilization, but a peculiar type of civilization, that leads to the communicating societies that SETI envisions. In fact, it seems that Western-style science may be critical to the entire process.

F_L: F_L is the final factor to consider. It requires estimating the lifetime of technical civilizations relative to some cosmic time scale, such as the lifetime of their own planet or the galaxy as a whole. In the alternative versions of the equation, the lifetime of the civilization (L) alone is used.[23] If technical civilizations mismanage their technologies and destroy themselves, their lifetimes may be very short. If they learn to use their powers wisely, they may live for millennia.

In terms of the equation, this factor can be understood as that fraction of civilizations that have survived and are still in the communicating phase.

On Earth, the lifetime has been only a few hundred years so far, and we don't know if our civilization will survive, or continue its interest in communicating. We have little guidance as to whether to assign a high or low percentage to this factor.

The mushroom cloud rising above Hiroshima and Nagasaki clearly warned humanity of our potential extinction twelve years before *Sputnik* soared into space and pointed to our next step in evolution. The planet has managed to avoid nuclear destruction over the past fifty years, but ecological hazards remain a clear threat. Some scientists studying the rapid extinction of terrestrial species today compare it with the time of the great dinosaurs.

Their work should sound an alarm for humans, because these species may be necessary to our own survival in ways we do not yet understand.

Human activities such as the burning of the Amazonian rain forest destroy animal and plant species on a weekly basis, and many people are trying to prevent those activities simply to preserve life itself. It may also be in our self-interest to stop the killing, because humans may become the victims as well as the victimizers.

Nuclear war can be prevented by international arms control agreements, and very optimistic trends toward peace and planetary unity are emerging as we approach the end of the century. We do not yet know, however, how to resolve the conflict between economic growth based on exploitation of natural resources and ecological balance based on respect for nature.

Current ecological problems have appeared when only a portion of the earth's societies are living in what would be called a "modern industrial society." As less developed countries modernize, the problems are magnified. Even without modernization, the pressure on the environment increases with population pressures, as in Brazil.

The space development community has proposed one solution, which is to become a multi-planet species. Critics say that if the same mentality is projected into the solar system as has prevailed on Earth, the same problems will arise. An alternative approach is to abandon the path of a high technology civilization, and return to a "low-tech" tribal approach.

Have extraterrestrial communities faced this problem and backed away from expansion? If so, they probably aren't communicating or listening. It takes not only advanced technology to do that, but also a surplus of funds and intellectual energies.

If the Assumption of Mediocrity is applicable to social evolution, it means that the technologies for interstellar communication and transportation arise simultaneously with the means of extinction, and the universe may well be littered with the remnants of civilizations that failed to resolve this dilemma.

Civilizations might reach this dangerous moment in their histories and find the answer precisely because intelligence *is* adaptive and helps them see how they must resolve the complex issues facing them. If so, that would suggest a vast number of civilizations might be in the communicative stage, especially if new societies learn from the old what they must do to survive.

Finally, the number of communicating societies might diminish if it turns out that further social evolution reduces the need for, or interest in, reaching outward to make contact with other species. Perhaps many technological civilizations survive the "danger period," only to abandon technical civilization itself.

By setting this factor at a high percentage, we are implying, then, not only that the lifetimes of technological civilizations are relatively long, but also that the communicative phase continues over time. A low percentage implies either that the lifetimes are short, or that the communicative period is brief.

Using the Drake Equation: "We Are Not Alone"

The Drake Equation moves from left to right, from the known to the unknown, and measures the limits of human knowledge. Putting numbers into the model can be misleading because it may suggest a better understanding of the situation than really exists.

However, it is an extremely useful tool for illustrating how a number of "assumptions of mediocrity" can produce many communicating civilizations, while variations on the "anthropic principle" can produce a few, or even one. Let's now go through that exercise and see how it works in both cases.

Sagan and Shklovskii calculated in the mid-1960s that there might be fifty thousand to one million communicating civilizations in our galaxy alone.[24] In our analysis, the result will be much lower, but still high enough to be tantalizing. We will be generating lower numbers than Sagan and Shklovskii because the assumptions will be far more conservative.

Filling in the equation begins with an estimate of the number of stars that might have planets circling them that are fit for life. Current estimates for the total number of stars in the galaxy range from 200 to 400 billion, providing an enormous population with which to begin the calculations.[25] However, it can be made more realistic by using only the number of solar-type stars in the galaxy.

Estimates of the number of solar-type stars in the galaxy vary from about one billion to nearly ten billion. In this first example, one billion is used, and the equation now reads:

Number of civilizations = 1,000,000,000 × F_p × N_e × F_1 × F_i × F_c × F_L

The second factor to be considered is F_p, or the fraction of stars with planets. Using the work cited earlier, it seems valid to assume that as many as fifty percent of all stars may have one or more planetary companions.

Using fifty percent yields an equation that reads:

Number of civilizations = 1,000,000,000 × .5 × N_e × F_1 × F_i × F_c × F_L

Multiplying out the two figures gives us 500 million (500,000,000) stars in the galaxy with planets—quite a large number, but one that is based on solid research.

Step three is to estimate how many of the star systems produce planets that are ecologically suitable for life. Unlike the first two factors, where some external information is available, this estimate must be based upon a sample of one, our own solar system.

Using our solar system as a model, there might be as many as thirty planets and moons in any given system. However, the requisite conditions for life do not exist on all the planets and moons of the solar system. Let's assign the number three to this part of the model, using the Earth, Mars and Titan as justification, even though Titan is outside the zone of habitability.

With three for N_e, the equation now reads:

Number of civilizations = 1,000,000,000 × .5 × 3 × F_1 × F_i × F_c × F_L

Having divided the sample in two (one billion times .5), it has now been multiplied to a higher number (three times 500 million), and the overall sample stands at 1.5 billion (1,500,000,000).

However, just because a planet can potentially evolve life does not mean that it does so. In the solar system, only one third of the potential life sites, or one (Earth) actually supports life now. Using .33 for F_1 results in a model that reads:

Number of civilizations = 1,000,000,000 × .5 × 3 × .33 × F_1 × F_c × F_L

The number of candidate planets with a foundation of living systems now equals about 500 million.

The next question is, "Of all these planets with living systems evolving on them, how many will go on to develop intelligent life?" Our earlier discussion shows that this question may be the "great divide" between abundance and scarcity in the universe; it is a difficult one to answer.

Given the uncertainty surrounding this issue, let's assign a relatively low number to F_i, calling it .01, or one percent. The equation now reads:

Number of civilizations = 1,000,000,000 × .5 × 3 × .33 × .01 × F_c × F_L

With this reduction, the number of planets with intelligent life is down to five million (5,000,000).

Five million planets with human-type life on them is an intriguingly high number, given the conservative assumptions that generated this answer. However, we still must decide how many of those planets produce technological civilizations. Since the evolution of a technical civilization on the platform of an intelligent life form is clearly not inevitable, let's again use .01, or one percent, as the number of planets that produce a technical, communicating civilization.

The model now reads:

Number of civilizations = 1,000,000,000 × .5 × 3 × .33 × .01 × .01 × F_L

The projected number of technical civilizations capable of interstellar communication is now down to fifty thousand, still rather a large number.

The final factor to consider is the lifetime of those technical civilizations capable of interstellar communications.

The final number is the most important and the most speculative of all. In the only known case, the issue still hangs in the balance. If it is assumed that the lifetime of such a civilization is relatively long in most cases, the number might be set at fifty percent, or 25,000 civilizations, or even ten percent, with a result of 5,000 communicating civilizations.

At one percent, the final answer goes down to 500. In this case, let's stay with the most conservative figures all the way, and use one percent.

The Drake Equation now reads:

Number of civilizations = 1,000,000,000 × .5 × 3 × .33 × .01 × .01 × .01
Number of extant communicating civilizations = 500

The most interesting question of all may be—are five hundred advanced technological civilizations in the galaxy many or a few? The answer probably depends upon our expectations. Considering the rich diversity of social and political life on Earth, with fewer than two hundred nation-states, five hundred planetary cultures should be enough to form the foundation of a fascinating galactic civilization.

Also, if the same numbers were to hold in the billions of other galaxies of the universe, it would mean that there are many billions of civilizations available to participate in interstellar and intergalactic communication. These strikingly large numbers form the underpinning of the Assumption of Mediocrity, and it is difficult to consider such quantities without concluding, "They must be out there."

The model's unknowns make it an uncertain foundation for predicting the probabilities of success for a SETI search. Its value lies in helping us to learn about SETI and think more clearly about it by seeing the impact of different assumptions. As new data is brought to light, it can be entered into the equation, making it far more accurate. Existing assumptions can also be tested to conduct a "sensitivity analysis" for impact on results.

For example, consider what happens if research on planets outside the solar system creates less optimistic numbers, and only 20% of the stars have planetary companions. Then the equation would read:

Number of civilizations = 1,000,000,000 × .2 × 3 × .33 ×
.01 × .01 × .01
N = 200

The answer drops to 200 civilizations.

Going in the other direction, if the number of solar-type stars were increased to two billion, the equation would read:

Number of civilizations = 2,000,000,000 × .5 × 3 × .33 ×
.01 × .01 × .01
N = 1000

The number of civilizations leaps to 1000.

Assuming that technological civilizations live longer also changes the outcome. For example, if we take the original equation, and change only the last value from .01 (1 percent) to .05 (5 percent), the result is:

*Number of civilizations = 1,000,000,000 × .5 × 3 × .33 ×
.01 × .01 × .05*
N = 2500

Now the number of civilizations escalates to 2500.

As a rule, the Drake Equation has not been used to conduct sensitivity analyses. However, analyzing a large number of cases reveals which variables are the most important. Changing variables also helps us to understand the different schools of thought more completely.

Using the Drake Equation: "We Are Alone"

Can you imagine a universe structured so that we are indeed alone as an advanced technological species? The Drake Equation can show the ways in which this might be so.

The Drake Equation reveals a variety of situations resulting in our being the only extant communicating species right now. The model shows that it is possible to get an answer of "1" *without* assuming that there have never been such species before. It is also possible to get an answer of "1" by assuming that life is common in the galaxy, but not intelligence. It is even possible to assume an abundance of intelligence, with only one communicating civilization.

For example, the equation might look like this:

$$N = N_s \times F_p \times N_e \times F_1 \times F_i \times F_c \times F_L$$
$$N = 1,000,000,000 \times .1 \times 1 \times .01 \times .01 \times .01 \times .01$$
$$N = 1$$

In this example, the one billion solar-type stars remain the same, but only 10 percent, or 100 million, are accompanied by planets. In those cases where planets exist, only one per system is ecologically suitable for life to arise. One percent of those succeed in supporting life's beginnings, or one million.

Of those one million, one percent (10,000) produce intelligent life forms, and of those, only one percent (100) produce communicating civilizations. Finally, of those 100, only one percent (one) are in existence today, i.e., Earth.

The example does not show that human beings are alone in the sense that the galaxy is barren of all life and intelligence except on Earth. It says that we are alone as a species at our particular level of evolution.

The scenario assumes a million planets blessed by life, and 10,000 with intelligent life. There are even one hundred advanced civilizations that have come into existence, but ninety-nine of them have either disappeared or stopped communicating. Perhaps they did not resolve the dilemma of technology growth versus ecological preservation, and perished. Perhaps they resolved it by turning back to a low-tech lifestyle. They may have become a spacefaring species, but are putting all their energies into building a Type II civilization, rather than into interstellar communication.

The equation can also state the most extreme of all possible cases, in which our sun is the only star with planets, and Earth is the only planet with life and intelligence. That might look like this:

$$N = 1,000,000,000 \times .0000000001 \times 1 \times 1 \times 1 \times 1 \times 1$$
$$N = 1$$

In this example, there are still one billion solar-type stars, but only one billionth (.0000000001) of them produces planets. Of those, only one (Earth) is ecologically suitable for life. Once that one appears, it alone produces life, intelligence, and an advanced communicating civilization.

Extending the Equation

The galaxy is enormous, but it is still only one of perhaps 100 billion such star systems in the universe. The stars of the Milky Way galaxy and those of other galaxies appear to be very similar; if our assumptions about evolution in our galaxy are correct, they are likely to be extendable to the universe as a whole.

Thus, if a conservative estimate suggests that there are some five hundred advanced civilizations in the Milky Way galaxy, there might be as many as 50,000 billion (100 billion times 500) advanced civilizations in the universe.

Even using the most pessimistic estimates regarding our own

galaxy, the probabilities expand again when the entire universe is considered. If it requires the mass and energy of an entire galaxy to produce one advanced technical civilization, there might then be 100 billion such civilizations in the universe.

There is no reason not to extend the equation to other star systems other than the fact that the distances (and therefore the time lag in communication) are much greater than with stars in the Milky Way galaxy.

Back to Probabilities

Now that the Drake Equation is fully understood, how can it be used to address the questions surrounding probability of contact?

Probabilities are estimates of how likely it is that an event will occur. A one-hundred-percent probability means it is certain that the event will occur; a zero-percent probability means the event certainly will not occur.

In the case of SETI, the probability of success is directly proportional to two critical factors: (1) the number of communicating civilizations; and (2) the amount of search space that is covered. As the number of civilizations increases, so does the probability of success, regardless of the amount of search space covered. Similarly, as the amount of search space goes up, so does the probability of success, regardless of the number of civilizations. If both quantities increase, so does the probability of contact.

For example, if one hundred percent of all possible stars harbor communicating civilizations, and we scan only one percent of them, the search will be successful. On the other hand, if only one percent of the stars have civilizations, but we cover all of the search space, we will also succeed.

Human beings on Earth can control only one variable, which is the amount of search space covered. The problem is that "search space" is not just a physical quantity. The term encompasses wavelengths and frequencies, and is also affected by length and sensitivity of observation. The truth is that we cannot hope to cover all of the search space defined in that way.

Humans have no control over, and little knowledge about, the other variable, which is the number of communicating civilizations.

All that can be said for certain is that as either of the variables increases, the probability of contact increases as well.

There was a time when large areas of planet Earth were unknown to mapmakers. They accurately depicted what they knew, and peopled the mystery regions with strange creatures of their imaginations. Today, the Drake Equation is a map of both knowledge and mystery for humans. The first few variables of the equation are the regions of the known, and are the domain of astronomers and astrophysicists. Farther to the right lie the regions of the unknown, populated by science fiction writers, and by the *Contact* conferences and the Cultures of the Imagination simulations.

Over the centuries humans will transform the Drake Equation from a model filled with assumptions to a description of reality in our galaxy and the universe. If space exploration in general is seen as a search for extraterrestrial intelligence, then the Drake Equation might well be seen as the master project-plan.

Someday our descendants will fill in the variables of the equation, as we learn more about the solar system, galaxy and the universe. Somewhere, on some planet in another star system, the answers may already be known. Of course, the best way to fill in the blanks is with continued research, such as the SETI projects that are underway and on the drawing boards.

Chapter Seven

Contact Scenarios

If we make contact, it will open unimaginable horizons. . . .If there is no contact . . . then that fact will have equally traumatic ramifications. . . .

—Gerald S. Hawkins,
Mindsteps to the Cosmos

DURING THE PAST THIRTY YEARS, RESEARCHERS HAVE CREATED THE tools necessary to conduct a systematic scientific search for extraterrestrial intelligence. The physical science of the effort is in its adolescence; the *social* science is in its infancy. The time has come to ask, "How is terrestrial society reacting to the search, and how will it react to successful detection or contact?"

Society has already felt the impact of the SETI Factor, especially in the past thirty years. Public knowledge of the scientific search has been minimal until now, but that will change with increasing commitment to it. We now need to prepare to cope with an empty universe, or a universe teeming with other species.

The previous six chapters provide enough information to begin building several detailed scenarios exploring what might happen in the SETI field over the next quarter-century. These scenarios are themselves forms of exploration rather than prediction, and are the next steps beyond the Drake Equation.

The scenarios take the information at hand and follow it through to logical conclusions. In this chapter, we will lay out options for the next quarter-century (1990–2015), a time period long enough to offer room for some of the major SETI projects to become operational and yield results, yet short enough not to extend into areas of pure speculation.

Readers of these scenarios should not think of them as final or complete, but as a starting point. In addition, SETI researchers ought to begin developing new scenarios so that we can anticipate the best response if and when one is called for.

The analysis is divided into "meta-scenarios," three general categories of possibility, and "scenarios," more specific themes within each general category. As this approach is refined, we can develop "micro-scenarios, which are still more specific and are discussed later in this chapter and in Chapter Eight.

Here, then, are three "meta-scenarios" that might play out over the next twenty-five years:

1. *No Contact:* SETI projects have been operating for thirty years now, without confirmed contact. Even with far more sophisticated techniques available, there may be no detection or contact in the next twenty-five years.
2. *Contact Within the Milky Way Galaxy:* If contact is made, it seems likely that it will be made with extraterrestrial intelligence within our galaxy, partly because that is where most of the search resources are going.
3. *Contact With Another Galaxy:* While contact within the Milky Way seems most probable, efforts to search other galaxies may also produce results.

Let's now look at each of these options, and the scenarios within them, in more detail.

No Contact

The most obvious reason for "no contact" is that no one is there to be contacted. This is the extreme version of the Anthropic Principle, in which ours is the only solar system in the galaxy and/or universe to have produced life and intelligence.

The "No Contact" option does not necessarily mean only that. It says that, for whatever reason, human beings do not contact extraterrestrial intelligence within the next twenty-five years. The following scenarios detail some of the varying reasons for this outcome.

Scenario 1-A—No contact because no one is there: This scenario affirms that those who believe in the utter uniqueness of humanity are correct. We would undoubtedly begin to speculate as to why that would be so. Perhaps the supporters of the Anthropic Principle are correct, or perhaps a few civilizations did emerge, and then died out.

It may be that the widespread distribution of life and intelligence throughout the universe is still the evolutionary strategy of the universe, but it requires a single planet as the source-point. Such a strategy could be adaptive, because it might spread life and intelligence throughout the universe more rapidly than having many source-points.

If contact is not always peaceful, but is more like the contact between cultures on this planet, the result could be protracted conflict between civilizations of different star systems. If the universe is programmed to evolve toward greater levels of life and intelligence, that would slow the process considerably.

A human species evolving outward from a single mother planet might be able to maintain a sense of unity and shared purpose over longer periods of time, serving the universal purpose more swiftly.

If no one *is* there, the probability of successful contact is absolutely zero, no matter what human beings do about it—the answer to the question "Are we alone?" is yes. Not enough searching will have been done to confirm this conclusion by the year 2015. However, the conclusion that no one is there will become far more plausible by then if major SETI efforts have failed to produce any results.

Scenario 1-B—No contact because they are beyond the search space we are exploring: The eight hundred to one thousand stars of the targeted NASA project seem like a lot, but they may not harbor extraterrestrial intelligence or advanced civilizations. An entire universe may not be needed to evolve one species like ourselves, but it may require large volumes of space within a galaxy to do so. The Drake Equation obligingly produces answers ranging from one to a billion civilizations. If the real answer is ten or twenty within the Milky Way galaxy, that's still exciting, but it will be harder to find them.

All of the searches may be overlooking the right areas of signal origin in ways that are not yet understood. There may be communicating civilizations within the galaxy, but humanity is just missing them. Alternatively, humans may be the only civilization in our galaxy. There may be others in different galaxies, but we will have to spend much more time looking for them.

The failures might persist over the next twenty-five years, and then success could occur if the resources invested in the process increase. The question is whether humans will continue to invest in SETI if there are no concrete results.

Scenario 1-C—No contact because the techniques are wrong: If someone were trying to call you on the telephone, but you were standing outside looking for them to drive up to your house, you might miss their call. All of the theories about the most logical way for extraterrestrials to communicate may be wrong. Perhaps finding the right way to listen is the ticket for entry into galactic civilization. Perhaps a new society is not deemed fit for membership until they have overcome their fears and become "senders" as well as "receivers."

Methods of communication have changed rapidly on Earth. In the past hundred years, humans have gone from letters and telegrams to electronic mail and telephone calls, with television and radio having appeared during that time as well.

The extraterrestrials may have invented a far better method of interstellar communication, and they may have abandoned the electromagnetic spectrum centuries ago. Would they leave behind a "beacon" for evolutionary laggards? Do twentieth-century humans maintain telegraph lines just in case someone discovers that technology a bit late in life?

If this scenario is accurate, failure might dog the SETI effort for quite some time, but contact could still happen if SETI researchers persevere and funds remain available.

Scenario 1-D—No contact because they are not in a communicative phase: This scenario is slightly different from the previous one. Rather than being too advanced in communications technology, the extraterrestrials may be too far behind us right now, or may have chosen different paths for social evolution. Current SETI projects, confined to a listening mode, can only receive a signal from a civilization technologically equal to, or more advanced than, human civilization. We cannot hear from civilizations evolving along an Earth-like track that have not yet reached a technical phase.

Since so little is known about alien evolution, we must also accept the possibility of advanced cultures without any interest in our form of communication. Lynn Harper suggests that different extraterrestrial cultures might evolve along radically different lines, as cultures on Earth have. Some might be obsessed with interstellar communication, investing vast resources in such projects; others might be introspective, ignoring the external universe, while cultivating a deeply meaningful inner life.[1]

We tend to use the term "advanced" in reference to technological and material matters, but there are forms of social and psychological advancement quite different from that narrow perspective. It may be that, on planet after planet, the "SETI phase" of a culture's history is brief and eventually abandoned. Imagine a culture several thousand years older than ours that has gone through an intense communications phase, and then evolved beyond it. Rather than having chosen a different path, they have transcended the path humans are now treading.

As in Scenario 1-A, no amount of effort on the part of SETI researchers will yield success if this scenario is correct. Nothing short of dramatic changes in detection methods would find these beings.

Scenario 1-E—No contact because they are avoiding us: Many SETI advocates favor this scenario as a way of resolving the Fermi's Paradox. It is a variation of the "Prime Directive" of "Star Trek" fame, which forbids the crews of Federation starships from interfering in the evolution of any culture discovered during their explorations.

The "Zoo Hypothesis" is another version of that scenario. It is the notion that humans are in a kind of "cosmic zoo," and, in the words of author Eugene Mallove, "Advanced cultures do not violate our territory, either because they fear rude and dangerous adolescent behavior, or they simply want to observe pristine biocultural evolution without disturbing it."[2]

Dr. Michael Papagiannis of Boston University tends to agree with this type of explanation for the absence of contact so far:

I have the feeling that there might be some kind of ethic in the galaxy that you don't contact new civilizations until they prove that they are really capable of overcoming their problems and establishing themselves as a peaceful and stable world.[3]

No Contact: Summary

Lack of success during the next twenty-five years tells us nothing in and of itself, except that contact is harder than we might have thought it to be. In two of the scenarios, terrestrials are using inadequate search techniques; improving the techniques would increase the chances of success. In three of the scenarios, the problems are connected with the extraterrestrials, whose existence and/or behavior is beyond human control.

A combination of circumstances might also cause the lack of success. The galaxy might be populated by some extraterrestrials using frequencies unknown to us, others using technologies unknown to us, others who have chosen an inward path, and still others who are waiting until we have evolved further before letting us know they are there.

If twenty-five years of searching does not produce contact, how long will human beings invest resources in a project without tangible results? If space exploration and settlement go forward, they might produce detection, contact, and/or encounters without conscious intention. However, if intentional searches are abandoned, a great opportunity will be lost. That's why thinking about "no-contact" scenarios is important.

Let's now look at the possibilities of success, starting with contact within our own galaxy.

Contact Within the Milky Way Galaxy

Most of the SETI researchers' attention is directed toward the Milky Way galaxy, and if contact is made with extraterrestrial intelligence in the next quarter-century, it probably will come from a star system within the galaxy. What are some of the possibilities that might be waiting out there now?

Scenario 2-A—Contact within the solar system: Contact within the solar system, even on Earth, cannot be ruled out, even though SETI scientists discount the possibility.

Contact within the solar system might take many forms. Astronauts could discover unmistakable signs of extraterrestrial visitations to some planet or moon, for example. The speculative notions of unconventional life forms, such as balloon creatures floating in Jupiter's atmosphere, could prove to be right, after all.

Explorers might find something like the monolith from Arthur C. Clarke's science fiction classic, *2001*, directing humans to take the next step in the cosmic journey of evolution. A true encounter would be the most dramatic event—a landing of extraterrestrial spaceships, in the open, without ambiguity, clearly making their intentions known to us.

Scenario 2-B—Contact with a nearby civilization: If the NASA targeted survey succeeds, it will produce contact with a nearby civilization. This scenario is most tantalizing, because it is a realistic possibility, and so much is known about it. Some of the best scientists on Earth are using some of the most advanced technology available to look at up to one thousand sun-like stars within 80–100 light-years of Earth.

Nearby civilizations also have a reason to aim their beacons at Earth. Members of the NASA SETI project have reported that the first commercial television signal broadcast from Earth may have gone out from Chicago's Channel Two in August, 1940. By 1950, that signal was ten light-years away from Earth, and had reached three solar-type stars.

Of those three, two are close enough to have received the signal and responded to it within the period 1940–1950. By 1960, the signal had passed eleven stars like the sun, but only the two were close enough to have responded in those first twenty years.

By 1980, the signal had passed ninety-two solar-type stars, and twenty-one could have responded in that forty-year time period. By 1992, the year in which operations for the NASA SETI project will begin, 156 solar-type stars will have been reached by the signal and forty-eight will have had time to respond.[4]

The Channel Two signal is being followed into space by an

enormous volume of signals that are expanding outward as a kind of cosmic "communisphere" from Earth, announcing the human presence to the local stellar neighborhood. If anyone is eavesdropping out there, this scenario can become a reality very soon.

This scenario is also fascinating because dialogue is possible within a reasonable time period. In the case of a star ten light-years away, several messages could be exchanged within the lifetimes of many who are alive when the first contact is made.

Scenario 2-C—Contact with a distant star system: Most stars in the Milky Way galaxy are distant from Earth. The sun is 27,000 light-years from the center of the galaxy, and the diameter of the galaxy is about 100,000 light-years. This means that it is about 50,000 light-years from the center to the farthest reaches of the system. If we look away from the center and outward from our position to the outer edge, the most distant stars will be 23,000 light-years away (50,000 minus 27,000). If we look inward to the center of the galaxy and then beyond to the outer edge in the other direction, the distance is 77,000 light-years (27,000 to the center and then 50,000 out again).

Within the galaxy, signals could be received from civilizations as close as four light-years, but as far away as 77,000! Dialogue is possible for closer stars, but we might also receive a signal sent 77,000 years ago, requiring a roundtrip of 154,000 years!

Real dialogue breaks down long before we consider communicating with the edges of the galaxy, however. True communication between our civilization and one that sent a message twenty thousand years ago is as unlikely as with a civilization that sent its message fifty thousand years ago.

These civilizations may have become extinct, or evolved to a point where they are no longer on their planet of origin. Beyond a few hundred light-years, the discussion is more about detection than two-way communication.

One-way communications are not meaningless in terms of impact, however. For example, if Civilization "A" simply sends a "hello" signal to Civilization "B," and then has to wait one thousand years for a response, it will take a long time for anything meaningful to happen.

If Civilization "A" sent a hello signal followed by all of that civilization's knowledge, it would be quite a different matter. Sagan and Shklovskii have calculated that all human knowledge could be sent out from the planet in four days. It would still take hundreds or thousands of years to arrive, but then all of the information would be received by the other civilization in a few days. That would be comparable to discovering a vast hidden storehouse of scrolls from a dead terrestrial civilization, such as ancient Greece or Rome. It would not be possible to have a two-way conversation with them, but the discovery would be a treasure, nevertheless.

If a now-dead civilization at the farthest reaches of the Milky Way were to send all of its knowledge, it would be of inestimable value, even though we could never discuss it with them. The actress Lili Tomlin, serving as hostess on a recent "Nova" program about SETI, said that she has "one-way conversations" with people like Socrates and Shakespeare by reading their works, and that these are valuable experiences, even though she cannot actually converse with them.[5] Contact with, or detection of, an extinct distant civilization would be very much like that, especially if they were wise enough to send the works of their Socrates and their Shakespeare!

Scenario 2-D—Contact with a galactic civilization: Most "SETI scenarios" focus on receiving a signal from a single civilization orbiting a single star; the level of development is a planetary or solar civilization at best. If there are relatively few technical civilizations in the galaxy, this perspective may turn out to be correct. Earth could be entering the communicative phase simultaneously with several other societies, and galactic civilization may be quite young.

However, if galactic society is old and has produced many civilizations, then the first signal would more likely come from the whole galactic community. The communications from such a society would be less a search or a beacon, and more of a series of instructions, a how-to kit for participating in galactic society.

Dr. Michael Papagiannis has given considerable thought to this option:

> It makes sense to me that they would have divided the galaxy into jurisdictions . . . and each advanced civilization would supervise its own area and know what is happening in it in terms of the evolution

of different planetary systems. And if each one of them covers a part of the galaxy up to at least one hundred light years around it, it is very possible that, with a super-advanced technology, they will know which stars have planets, and which of those planets have life. . . . They would keep an eye on those particular places rather than randomly sweeping the whole universe with their antennas.[6]

Papagiannis suggests that these "galactic watchers" might also send unmanned probes to a developing solar system to provide advanced surveillance capabilities. The galactic society would not reveal itself until the evolving civilization was judged to be ready for it, and only then would the signal be sent.

Contact Within the Galaxy: Summary

The primary distinction of the first two "meta-scenarios" is between contact and no-contact. When contact is considered, distance is critical, because it structures the nature of the contact. At great distances, detection or one-way communication is all we can hope for, unless there are technologies available to overcome the barrier of the speed of light.

The issue of single/multiple culture contact is also significant. Just as the difference between close and far is qualitative as well as quantitative, so is contact with a full-fledged galactic civilization quite different from a dialogue with another society struggling to find its place in the immensity of the cosmos.

Contact With Other Galaxies

To Earthlings, distances within the solar system are immense. Humans contemplating manned missions to Mars wonder whether people can handle a journey of several months across the gulf between Earth and the red planet. Yet, Mars is one of the closest planets to Earth within the entire solar system.

The nearest star, Alpha Centauri, is four light-years away—so far that even if human technology could accelerate a spaceship to lightspeed, it would take four years to get there. The galaxy is 100,000 light-years in diameter—it takes a ray of light one thousand centuries to cross it.

What, then, of the other galaxies and any hopes for contact with extraterrestrial intelligence there?

It appears that the same basic conditions apply in other galaxies as in the Milky Way. The fundamental constituents—stars—are the same, and they seem to be made of the same materials. Galaxies are not evenly distributed throughout the universe, but are clumped together into "clusters" and "super-clusters." Groups of galaxies are linked to one another in much the same way as planets in a solar system, by the mutual attraction of gravity.

The Milky Way belongs to a "Local Group" of some thirty galaxies. These nearby galaxies include the Large and Small Magellanic Clouds, and the enormous spiral galaxy of Andromeda. The "Local Group" is local in name only: the Earth is 170 thousand light-years from the Large Magellanic Cloud, and 190 thousand light-years from the Small Cloud. Andromeda, by comparison, is over a million light-years away!

Moving beyond the Local Group of galaxies means that any other galaxy to be considered by a SETI program will be millions of light-years away, with all that this implies for contact.

Scenario 3-A—Contact with a civilization within a Local Group galaxy: Just as SETI researchers on Earth are looking at nearby galaxies for sources of intelligent signals, a communicating civilization might send a signal to another galaxy within the Local Group. Because of the distance and time involved, any signals would be like a message from a very distant star within the Milky Way—dialogue would not be possible within our own lifetimes, though enormous amounts of information might be received in a brief period of time.

Scenario 3-B—Contact with a civilization from a galaxy outside the Local Group: This scenario represents the outer limits of probability, since the distances between the Milky Way and other galaxies may be millions or even billions of light-years. It is included as a theoretical possibility because a signal from such a source does answer the question, "Are we alone?" which is reward enough for the effort.

Scenario 3-C—Contact with a universal civilization: If planets evolve civilizations composed of different cultures, galaxies can evolve civilizations composed of different planets, and galaxies may form inter-galactic, or universal civilizations.

In order to exist, a universal civilization will have overcome the distance barriers to communication, and would know how to create a message that would be useful to those receiving it in other galaxies. Considering this scenario to be real is difficult, and the implications for human beings are almost beyond comprehension. But the possibility that the universe has evolved so far is also thrilling to contemplate.

Summarizing the Scenarios

We have now created and described twelve scenarios that are plausible outcomes of the SETI search over the next twenty-five years. They cannot be all-inclusive, but represent a beginning for social scientists who want to take the next step in preparing for the impact of SETI.

The options range from an impressive silence ("No contact because they are not there") to a truly stunning announcement ("Contact with a universal civilization"). The scenarios move the discussion out of the simple contact/no contact domain, and the real importance of SETI emerges. Whatever the answer, every scenario says something fascinating about the universe.

Any one of the scenarios could become reality. The point is not to find the "right" one, but to use them to prepare for the actual event—just as astronauts spend hours in simulators so they won't find the experience of outer space too disconcerting.

These scenarios provide a practical way to "get ready for SETI" by exploring the impact on Earth resulting from any one of them. That in turn opens up the possibility to shape the evolution of the response, and therefore of human society, rather than passively reacting. We can also use the process to create positive social changes, such as an increase in planetary unity, regardless of the outcome in reality.

Micro-Scenarios

Once the broad structure of contact scenarios is in place, creative minds will see the opportunity to go further. Why not imagine a specific star at a given distance from Earth, of a certain age and luminosity? Hypothesize a planetary system accompanying the star, a planet of specified mass, rotation period, revolution period, age, distance from its sun, and atmosphere.

A few simple assumptions lead to images of how life and intelligence might have evolved on such a planet and the state of its society relative to Earth. The final step in this process would be to simulate the encounter of these extraterrestrials with terrestrials, and see what happens.

It's such a good idea that it's been done. "Micro-scenarios" are the foundation of the science fiction genre; science fiction writers have long imagined such encounters, utilizing all our scenarios, and more, as the basis for their work.

A new form of "living science fiction" with major implications for SETI is being created now in the annual *Contact* conferences. *Contact* is the brainchild of anthropologist Jim Funaro, and it is a unique interdisciplinary approach to SETI.

Working with a select group of physical scientists, social scientists, aerospace experts, and leading science fiction writers, *Contact* conferences include several formats. One of these, called "Cultures of the Imagination" (COTI), provides participants with the opportunity to prepare for contact.

Using a process called "world-building," the participants begin with a set of assumptions about another solar system and proceed to develop detailed micro-scenarios. Participants divide into a Human Team and an Alien Team, working separately for the first few days of the event. Prior to the beginning of the conference, the groups are provided with assumptions about both the alien and human civilization, and they build their worlds from there, coming together only at the end of the conference for an unrehearsed encounter.

The process has produced a series of fascinating extraterrestrial species, including:

1983: The "Alchemists," who lived underwater and communicated primarily by smell. Also known affectionately as the "Flying Whales."
1984: The "Squitches," a bipedal herd species that laid eggs to reproduce and stamped their feet to communicate.
1985: The "Mossbacks," a reptilian species that evolved in the swampy lowlands of an extremely hostile world.

1987: The "Centaurs" of Alpha Centauri B, decapods named after the half-horse, half-human beings of terrestrial myths. Also known as the "Dragons."
1988: The "Evos," a powerful species of aliens working together as part of a group mind.
1989: The "Us," another group mind species; they ate or absorbed everything they encountered.[7]

Human society must also be defined at the time of the encounter, and this piece of the puzzle is developed with great care by the conference organizers. The encounters could generate a variety of outcomes because of the alien species, but varying the human situation guarantees that the results are even more unpredictable.

For example, the Centaurians are discovered by an unmanned probe that goes to Alpha Centauri, and is followed up by an expedition from Earth. By contrast, the Us stumble upon a human colony that has settled on a planet circling Tau Ceti, an outpost of a human galactic empire in decline.

Careful attention to scientific detail gives the alien species of *Contact* an air of reality that is essential for the success of the simulations. For example, at *Contact III* in 1985, the science fiction writer Poul Anderson, assisted by C. J. Cherryh and Larry Niven (all three being award winning science-fiction writers), created the planet "Ophelia." A brief description, based on the conference proceedings, follows:

Ophelia is assumed to be part of the planetary system of the star "Hamlet." Its mass is 1.75 times that of our sun, and its luminosity is 5.4 times as great. The light of Hamlet is said to be bright, "looking white, with a faint bluish tinge to the human eye, somewhat like a fluorescent lamp's."

Ophelia is about three times as far away from Hamlet as Earth is from our sun, and only gets about half as much radiation. Hamlet is younger than our sun, but because it is brighter, the star is evolving more quickly. Ophelia only has about half a billion years to live before Hamlet brightens significantly and triggers a runaway greenhouse effect.

Ophelia's diameter is about 1.8 times that of Earth, and its day is about sixteen hours. Its core is more massive and hotter than Earth's, and the rapid spinning of the core produces a strong magnetic field. The result: strong geological activity, including volcanoes, powerful tides, and variable weather. Five-sixths of Ophelia is water, but because the planet is larger than Earth, it has about 90% of Earth's land surface.

Looking at the harsh conditions on the planet, C. J. Cherryh imagined a life form evolving in Ophelia's swampy estuaries: ". . . being shaped liked a horned toad would be a real advantage in handling the swamps and spreading out for warmth, and riding the tidal waves without getting killed. Also, short and stumpy would do well with the higher gravity."[8]

Larry Niven offered his view of the intelligent "Mossbacks" that eventually evolve on this difficult planet: "Intelligent beings will almost certainly be air-breathers. They burn their food fast; they live fast; probably they breed fast, grow fast, die fast. If you want to talk to any intelligent Ophelian, you have to talk *fast*. . . ." Niven pictured the Mossback as having "a mucking great beak" with a short arm above it for tool-making.[9]

World-building is an exercise of the imagination, but it is rigorous, and world builders do not allow unfounded assumptions. For example, it would be unacceptable to describe life as we know it evolving on an airless planet a few million miles from an alien sun, unless there were good scientific explanations as to how that could happen.

Staying within the bounds of what is known about evolution of life and intelligence is not a limitation to creativity—it still produces exciting and stimulating contact scenarios resembling events that may take place in the future.

World-building is an excellent educational tool, helping participants learn more, not only about SETI, but also about themselves and interaction among cultures of any kind, including those on Earth. The founders of Contact now plan to introduce a "COTI Junior" process into school systems on a trial basis.[10]

The next stage in the world-building process might be "galaxy-building." If the exercise goes beyond one world to two, then three or four, simulating the interactions between them is the natural next step. Contact conferences have begun to do so by imagining an interaction between aliens and Earth. In the future, they can go further by having a simulation involving humans with several of the alien cultures created at earlier conferences.

As of 1989, the Contact conferences had not developed a scenario in which contact occurred on an electronic basis only, but that is another approach to consider. Micro-scenarios linked with the twelve scenarios developed earlier in this chapter can help to shape the impact and guide the process, regardless of the outcome with real SETI projects.

As the next chapter shows, humans can control only one variable in the SETI process, and that is our degree of preparation. Getting ready for SETI means taking control of our own evolution, a step that we simply cannot avoid taking.

Chapter Eight

Immediate Impact

It would be the biggest science story of all time.

—John Noble Wilford,
New York Times

THE HUMAN MIND NOW REACHES OUT FAR BEYOND THE PLANET OF its birth and seeks to understand the nature of the cosmos. Human consciousness is evolving by exploring the universe; asking questions about the universe shows our willingness to take major steps in evolution. The SETI question is clearly one of the most important ever asked.

What will be the impact of an answer, regardless of what it might be? Sometime over the next quarter-century, one of the three meta-scenarios described in Chapter Seven will be played out, and Earth will feel the impact.

Our immediate concern is how the answer will affect the human mind and civilization, but humanity is a subsystem of the whole system known as the universe—any change in a part will be felt by the whole. The impact is not limited; it is universal. We are confronting the question of how the universe itself is evolving, and where intelligence fits into that process. The question, "Are we alone?" must be re-stated as, "What is the evolutionary direction of the universe, and how many intelligent species are needed to take it there?" or "What is the extent of biological evolution and how often does intelligence appear?"

A clear and unambiguous communication from another star system will have the most immediate and direct impact on humans, but non-contact will affect us as well. SETI is getting greater publicity because of the proposed NASA search—more people want to know the outcome. If an intense period of SETI activity produces no clearcut results, the impact of non-contact will be real.

On the other hand, contact may occur, following one of the scenarios developed in the previous chapter. What will be the results of such contact, and how will the impact differ depending upon the scenario? Is there a "Drake Equation" of impact?

The Impact of Non-Contact

Starting in 1990, let's take 1992 as a milestone (the five hundredth anniversary of Columbus' voyage, the International Space Year, and start-up for NASA SETI Project operations) and then look ahead twenty-five years to 2015. Imagine the following micro-scenario:

> Announced with great fanfare in 1992, the NASA SETI project has continued for almost twenty-five years. It now includes scientists from the Soviet Union, the European Community, Brazil, Japan, and twenty other countries. Project META continues as well, with an expanded Southern Hemisphere and Northern Hemisphere search in operation.
>
> The "World SETI Federation" boasts membership of some 25,000 researchers and other interested parties. The United Nations has designated 2010–2020 as the "SETI Decade," and is investing major resources in the search.
>
> Interest in extraterrestrials remains high worldwide, and popular culture maintains an outpouring of "artificial reality" simulations, books, videos, and films that have turned "world-building" into a major art form.
>
> However there is still no proof. Almost sixty years after the paper by Morrison and Cocconi started it all, none of the planet's SETI programs has yielded an unambiguous signal clearly "ETI" in origin. The search has generated enormous amounts of data, and much has been learned about the universe. There are new theories about black holes and quasars, for example, but if the universe contains intelligent life beyond the earth, we still can't prove it.
>
> The situation is not much better within the solar system, which is being intensely explored by manned and unmanned vehicles. The human outpost on Mars has found tantalizing evidence of life having been there once—some even say there was intelligent life—but no one is certain. As expected, Titan seems to be in the earliest stages of evolving organic life, but that's it.
>
> UFOs continue to puzzle and plague the human mind, but remain

a matter of dispute, as they have for thousands of years. If they are the extraterrestrials we are looking for, they are a shy and subtle bunch!

The search for extraterrestrial life is more hopeful than the search for intelligence. Radial velocity studies have identified over one hundred stars with planetary companions, and optical telescopes in Earth orbit and on the moon are confirming these findings. Early results suggest that there are at least four planets with oxygen-rich atmospheres. However, when our radio telescopes are turned toward these planets, nothing out of the ordinary is heard, just the usual noise of the cosmos.

This micro-scenario based on the assumption of no contact may be wrong in its details, but it shows how it might look and feel if nothing definite happens within the SETI field for the next quarter-century.

The scientists are right when they say that the search is worth the time, energy and money even if it doesn't produce contact quickly. The search ought to continue, regardless of short-term outcomes, but human beings are not patient organisms. They lose interest in activities that go on and on without exciting and stimulating results—especially when money is involved.

While twenty-five, fifty, or one hundred years is an instant on a cosmic scale, it is a long time for a person or a society. If nothing amazing happens by the year 2015, it doesn't mean that SETI will be abandoned, but assumptions will be made, and policies will be enacted.

"We are alone" is an emotional statement, even if it results from a scientific experiment. There is something terribly sad in those words, especially when applied to an entire species and an entire universe.

Suppose most of the people living in 2015 conclude that Professor Tipler and the other doubters are right: there just aren't any extraterrestrials, and human beings must make their way into the galaxy and the universe on their own. The idea of "aloneness" will then take many forms. Religious people would point out that we are not really alone in the universe, because God, or the Universal Mind, is still there and always has been. Finding that there are no other intelligent life forms simply confirms religious views of humans as special and unique.

Others might say that we were never alone, anyway, whether we believe in a divine force or not. The millions of life forms on Earth, if we would only pay attention to them, are connected to us in subtle and substantial ways. We are part of a web of life that is mysterious and beautiful; we should spend more time searching for terrestrial intelligence, which includes not only ourselves, but the entire biosphere and perhaps the planet itself.

Finding that extraterrestrial intelligence, as defined by SETI, is scarce might well create a new and positive focus on the earth and its creatures. The work done in communicating with dolphins and chimpanzees is breathtaking in its implications, and renewed efforts in that area might provide us with much of what we are seeking elsewhere.

Native peoples say there was a time when humans talked with animals, but contact ended with the coming of the Europeans and the onslaught of technology. If we have made ourselves alone in the universe by cutting ourselves off from what is sacred, whether that be God or nature, or both, a disappointment in the SETI field might return us to our roots.

Still, to find that the answer is perhaps "yes, we are alone in terms of intelligent, communicating aliens being out there" may be difficult for our species to handle. The sadness may not be expressed or consciously confronted, but it will still exist, and we will need to find a way of responding to it.

Scientists will say that the lack of success doesn't prove anything at all, except that we should continue the experiment. That would be correct from a scientific perspective, regardless of the fact that it is not very satisfying emotionally and psychologically.

Policymakers will be concerned about investing more intellectual and financial resources in a project without clearcut benefits. Depending upon what else is demanding attention and funding, they may agree that SETI is valid "pure science," worthy of continued pursuit, or that it's a luxury we can afford no longer.

For science fiction writers and all those working in the realm of the imagination, the great pleasure of aliens is that "you never know." They might be there, or they might not. Condemning fairies, werewolves, and vampires to the realm of the imagination hasn't stopped writers and film-makers from using them, but it does change the nature of the portrayal. It is seen as fantasy, not as potential fact. The same is likely to happen with extraterrestrials if the culture decides they probably aren't out there.

Great science fiction has already been created without writing about extraterrestrials at all. In a series of books comprising a comprehensive "future history" of humanity, Isaac Asimov imagined a "no-contact" scenario in which human beings evolve a complex galactic civilization of thousands of worlds. This galactic empire is so old that it has forgotten its origins on a single planet known as "Earth."[1]

The books, including *Prelude To Foundation, Foundation, Foundation and Empire* and others, have been extremely popular, even though there is not an alien or extraterrestrial to be found anywhere in them. For the science fiction genre, it is an amazing feat to write so many books about a human-dominated galaxy.

Asimov even poses a theory to explain why we might be alone. In *Foundation and Earth*, he proposes that Earth's large satellite is the primary cause of its uniqueness. Conceding that evolution may occur in the same basic way everywhere in the galaxy, Asimov hypothesizes that it may move faster on planets with moons that are large compared to the planet itself, due to the gravitational influences of the satellite. In addition, the Earth's crust is slightly radioactive, which increases the number of mutations, and also speeds up evolution.[2]

Asimov's galaxy without extraterrestrials does not reflect his own views of what is out there.[3] The point is that science fiction writers will find a way to survive in a universe without extraterrestrials. But what about cosmologists? They will be directly affected by twenty-five years of intense, but unsuccessful searching. In particular, those who have advocated the Assumption of Mediocrity will be forced to re-think that position, since early returns would contradict their expectation that intelligence is ubiquitous in the universe.

Steven J. Dick believes that the cosmological impact of SETI is critical for all of us, because "We see the universe through the lens of our cosmology."[4] Like Gerald Hawkins, he views cosmology as evolving over time in concert with the evolution of human thought, and traces the historical development of six distinct cosmologies. These cosmologies are the following:

1. *The Atomists' Cosmology:* Developed by the Greek natural philosophers, this cosmology assumed that everything in the universe was made of tiny, discreet particles called atoms, which floated through space and came together to form worlds, on which there might be life and intelligence.
2. *Aristotle's Cosmology:* Aristotle saw the universe as composed of a series of solid spheres radiating outward from the earth and carrying the moon, sun and planets, with the stars rotating on the outermost sphere.
3. *Copernicus' Cosmology:* Copernicus' cosmology differed from that of Aristotle, in that the sun was placed at the center of the universe, rather than the Earth. This led to a number of new developments, especially Kepler's laws of planetary motion.
4. *Descartes' Cosmology:* Descartes' principal contribution came in the idea that the matter of the universe whirled around in a series of vortices, with the sun being at the center of one local vortex, while the other stars sat at the center of their vortices. There might well be life and civilizations that arose on planets revolving around distant stars.
5. *Newton's Cosmology:* Newton added order to the universe, with his mathematical expression of the laws of motion and gravity. Newton's universe is filled with galaxies, some of which may have solar systems, life and intelligence.
6. *The Bioastronomers' Cosmology:* This modern cosmology extends Newton's faith in a universe governed by universal laws throughout the cosmos and sees the building blocks of life everywhere in the universe. It also assumes that the laws of nature apply in much the same way throughout. Such a universe may be teeming with planets, life, intelligence and civilizations, some of which may be trying to contact us.[5]

Dick points out that the bioastronomers' cosmology has been dominant since around 1750, when the belief in a "plurality of worlds" achieved widespread acceptance. However, says Dick, while this cosmology has been *accepted*, it has never been proven. The critical importance of SETI, according to Dick, is that it provides "for the first time" a test for the dominant cosmology of our time.[6] If SETI is successful, it will confirm a cosmology already widely accepted. If it is not successful, the experiment will tend to contradict that cosmology, which may have to be replaced or revised.[7] This bears directly on the question of whether there are "privileged sites" in the universe. If the search succeeds, it will strongly suggest that Earth is not a privileged site in the universe, and that there probably are no privileged sites. However, if the search continues for a long period without success, it is likely that much more attention will be paid to the Anthropic approach.[8]

As we noted earlier, the differences in these two cosmologies reflect different views of humanity. The Assumption of Mediocrity, which underlies the bioastronomers' cosmology, inclines toward a relativistic view—there are many intelligent life forms, none is necessarily better than the other, and there is nothing special about human norms and values.

The Anthropic Principle leads to a more homocentric and absolute view, with human norms and values being extremely important to the universe as a whole.

The ordinary citizen will eventually experience the results of the SETI experiment, but subtly. Most people remain unconcerned with scientific SETI, though they express a deep curiosity about UFO reports. Very few even know that we have had thirty years of unsuccessful searching since 1959—they are unlikely to be aware of another quarter-century of non-contact.

One group deserves special mention, those who are most deeply involved in space exploration and development. Space activists are a significant subculture, having dedicated themselves to the dream of human expansion outward into the solar system and eventually to the stars. Their vision is defined as space development or space settlement, and is also known as interstellar migration. For some, it represents a kind of "cosmic human destiny," while for others, it is the necessary next step in human evolution. For still others, it is an insurance policy that will prevent human extinction, in the face of current threats to life and civilization on Earth.

Many scholars and scientists see benefits in opening up the "space frontier." It provides an opportunity to divert nationalistic energies away from war and toward peaceful cooperation ventures; it also offers an expanded range in which to work out new forms of social and political interaction. In The Overview Effect, I pointed out that space exploration also provides an opportunity for human awareness to evolve and transform itself because it provides us with a new perspective on the earth, the universe, and ourselves.[9]

The defining feature of the space development subculture is a refusal to consider the future of humanity as confined to the surface of one planet. While members of the space development community may be concerned about the future of Earth, it is not because they plan to stay here. They see themselves as the leaders in creating a "spacefaring civilization," and making humanity into a "multi-planet species."

Space development advocates historically have shown relatively little interest in SETI, though that has recently begun to change. Their vision is one of unfettered human expansion outward, and their interest is in escaping the bonds of Earth. Extraterrestrials are a problematic issue for space explorers, because we cannot predict the impact of their presence on the basic vision.[10]

It's not that space development advocates would welcome the news that humanity is alone in the universe, but it would have an effect on their vision, and provide a rationale for it. If the earth is the seed-planet of life and intelligence within this galaxy and beyond, then the human mission becomes clear: developing a strategy for accomplishing its purpose—universal dissemination of life and intelligence.[11]

As exciting as contact might be, the space development vision is also inspiring, and a universe without extraterrestrials need not be a boring place. It does mean, however, that the first extraterrestrials will be humans and their descendants!

Contact in Context

Scientists, journalists, academics and interested observers tend to agree that contact with extraterrestrial intelligence will have a significant impact on humanity, but details are scarce. If the deepest and most lasting impact will be on human consciousness, then we need to put the event into a historical context. Then, we can consider the impact of contact on the universe itself.

Gerald S. Hawkins has attempted to define a broad perspective for thinking about contact with extraterrestrial intelligence. He does so by viewing human history through our perceptions of the cosmos,

with the new understandings defined as "mindsteps." As successive
mindsteps occur, so do new perceptions, or cosmologies.

Hawkins provides a valuable way of thinking about intelligence
with his discussion of "memes." A meme is like a mental gene; it is
an idea useful for survival that is maintained by a culture over long
periods of time:

> Each little thought is called a "meme." A group of people share a set
> of memes. It is like sexual reproduction, where physical characteris-
> tics are transmitted by genes, and a cultural gene pool develops. . . . A
> gene pool interacting with the environment shapes the body. A meme
> pool interacting with the cosmos shapes the mind.[12]

At a given moment in history, the state of human awareness is
defined by the meme pool. According to Hawkins, mindsteps are
irreversible changes in human understanding of our relationship to
the cosmos that occur in conjunction with changes in how memes
are stored and manipulated. For example, drawing initiated the first
mindstep, defined as Mindstep Zero, that led from Neanderthal Man
to Cro Magnon Man, and began what Hawkins calls the Age of
Chaos. It lasted from 35,000 BC to 3,000 BC, when writing was
invented.[13]

Writing ushered in the Age of Myth and Legend, a time when the
universe was explained by stories of how the sun, moon and stars
were created. Hawkins argues that the separation of Heaven and
Earth was part of Mindstep One.[14]

Mindstep Two occurred when human beings discovered the or-
dering power of mathematics in the first centuries after Christ. In the
cosmological field, this age was defined by the Ptolemaic model of
the universe, which put the earth at the center, with everything else
revolving around it.

Mindstep Three came with the revolutionary reversal of the Ptolemaic perspective, as Copernicus argued that the sun should replace
the earth at the center of the solar system. Hawkins puts Mindstep
Four in the twentieth century, with the launching of the first liquid
fueled rocket by Robert Goddard, ushering in the dawn of the Space
Age.[15]

Each mindstep is linked with a new communications innovation,

Mindstep	Date	Age	Invention	View of Cosmos
0	35,000 BC	Age of Chaos	Drawing	Universe centered on the person
1	3,000 BC	Age of Myth & Legend	Writing	Earth and sky separated
2	150 AD	Age of Order	Mathematics	Earth the center of the universe
3	1543 AD	Age of Revolution	Printing	Universe centered on the sun
4	1926 AD	Age of Space	Liquid-fueled Rocket	Humans travel in space
5	2021 AD?	Age of Cosmic Connection?	Unknown	Communication with extraterrestrial civilizations?

Figure #1: Mindstep Model of SETI Contact (Adapted from *Mindsteps to the Cosmos*, By Gerald S. Hawkins)

a new age, and a new cosmology. According to Hawkins, there is initially a long period of absorption between the advent of a mindstep and the next mindstep. That period of absorption is becoming shorter, however, and mindsteps are occurring much more quickly in our time.[16]

Hawkins has developed an equation that predicts the next mindstep at around 2021, just beyond the quarter-century period used as a guideline for the analysis in this book.[17]

Nothing in Hawkins' theory points directly to the fifth mindstep being extraterrestrial contact, but he finds himself led in that direction:

Personally I feel the step will be in some way an extra-terrestrial one, and that tends to be the viewpoint of most astronomers . . . it seems to me that the slow adjustment of humans to the cosmos, through cave pictures, legends and telescopic viewing, has been leading toward a contact.[18]

Hawkins views contact with extraterrestrial intelligence as ranking with the most important events in human history: its impact can be compared to the Copernican revolution or the advent of the Space Age itself. Hawkins does not speculate on the details of how contact might affect Earth and human society. That work is left to others.

Building a New Perspective

Accepting Hawkins' view that contact would usher in a new age for terrestrials, how might we imagine the specific details of that era, and what can be done to make it a positive one?

The first step is to ask what is known about how human beings and societies process new information, because the announcement of contact will go out to a worldwide information processing system. As extraordinary as the announcement will be, we can learn a lot by analyzing how the system handles any form of news, and by what experts think the system would do with this news.

The second step is to examine the assumptions that human beings have already formed concerning contact with extraterrestrial intelligence. The assumptions may not be correct, but they show humanity's current state of awareness about contact, from which it may be possible to project changes over time.

The final step is to create a mental model like the Drake Equation, aimed at understanding the impact of contact—a Contact Impact Model. Like the Drake Equation, its value would be qualitative rather than quantitative, providing insight and education about how to handle contact.

The process must begin with what is known, moving on to what is unknown, as the Drake Equation does. In this case, however, the analysis is of social and psychological, rather than physical and biological, systems. The model must refine existing assumptions about contact, providing a solid new foundation for assessing impact.

Impact will depend on the type of contact being modeled. The scenarios in Chapter Seven range from extraterrestrials landing on Earth to a signal being received from a galaxy billions of light-years away. Common sense says that the impact will be different, but how different will it be? The model should address that question.

Let's begin by looking at how the world community processes new knowledge now.

The System

The planet Earth is an information processing system. In fact, the entire universe can be understood as an information processing system, with Earth functioning as one of many subsystems scattered throughout the cosmos.

Processing takes place at different levels, the sophistication of which depends upon the complexity of the system available. For example, any one of the billions of atoms making up the earth is an information processor of limited capacity. As atoms are combined into increasingly complex structures, such as molecules, cells, brains, animals, humans, and societies, they become capable of processing increasingly more complex information.

At the social level, humans have constructed an information processing system, a "technosphere," that is planetary in scope. It links most of the population with satellites, computers, television/radio networks, and print media. SETI is a relatively new component of this infrastructure, but it may be extremely important, and is itself triggering development of new information processing and storage capabilities.

The technosphere has in turn created a radically new environment for humans, which can be called the "communisphere." Increasingly, humans do not live in a natural environment, but in the artificial communisphere.

According to Ted Bonnitt, Editorial Director of *World Space Report*, the result is a homogenization of knowledge. He sees the system as delivering "factoids" to readers and viewers on a continuing basis. It's not that the information is untrue, but rather that it is abstract, and it's difficult to judge the relative importance of one bit of news compared with another.[19]

How does the system decide what to disseminate and how do people determine the value of that information? The decision is apparently made according to two criteria: one is the general "importance" of the news and the other is the immediate relevance to the individual. People judge the information primarily in terms of their self-interest.

Journalists want to pursue stories that are important but which also have immediate and ongoing interest to large numbers of viewers and readers. Consciously or unconsciously, media people direct the information processing system according to their own criteria and bias.

According to Kelly Beatty, Senior Editor of *Sky and Telescope* magazine, many stories are vital but "unfathomable" and difficult to explain to the public, because the stories have no immediate saliency to their lives.[20]

Journalists reporting on complex topics have to go to others for confirmation that their information is valid. They cover fields that are often beyond their own realms of expertise. Even a highly-trained science reporter can't hope to have equal knowledge of both space exploration and genetic engineering.

In reaction to major events, such as the announcement of the "cold fusion" breakthrough, journalists turn to opinion leaders in the scientific community for confirmation. In that case, detractors immediately expressed doubt regarding the announcement, and journalists who had been enthusiastic became cautious. As long as there were scientists who held out hope for the process, it continued to be covered. As the weight of opinion swung against cold fusion, the coverage died out.[21]

As Ted Bonnitt sees it, the problem begins with journalists as the first group of people to consider the implications of new information, such as the acquisition of a signal from another planet: "They will react like most people. . . . They will try to figure out what they have to lose in terms of credibility."[22]

David Baron, Science Reporter for WBUR-FM in Boston, points out that there is a reason why reporters seek out "big stories," but remain skeptical:

> The paradox is that the most newsworthy stories are the most surprising ones, and things that are surprising are often greeted with skepticism because no one expects it.[23]

News is an important media product, so journalists must seek out the extraordinary. At the same time, they want to avoid being duped, so they are pulled in opposite directions when the story is complex and the evidence difficult to interpret. Baron says that journalists have become much more cautious since the cold fusion controversy:

> That was a big story, with implications that were phenomenal, and that's why it was a big embarrassment to people who really pushed the story. . . . You can't help but have self-interest come into play. . . . No one wants to be caught reporting bad science.[24]

Baron and others note that a similar cycle—excitement contending with skepticism—also accompanied the reports on room-temperature superconductivity, an experiment eventually judged to be valid by the scientific community. In that case, once the validity was established, the interest diminished because after a while, nothing really "new" was announced.[25]

The system also gravitates toward extremes: the first, the last, the biggest, the smallest. After the first *Apollo* moon landing, the interest in the missions declined, and the same thing happened with the space shuttle until the *Challenger* accident.

To be sure, all journalistic operations are not the same, and there are cultural and credibility differences to consider. The reaction of state-controlled media to news events is not the same as the response of a private, or "free" press. For the "tabloid" or "trash media," being sensational is usually more important than being accurate.

The differences are apparent when stories about UFOs or the paranormal are the issue. For example, in the fall of 1989, *Tass*, the Soviet Union's official news agency, reported a landing by alien spacecraft in a town 300 miles from Moscow. The aliens were said to be tall, with small heads. They took a walk in the town's park, accompanied by a robot, and traumatized the citizens.[26]

Tass initially reported the landing as if it were true. Western news sources gave it quite a bit of play, but with tongue-in-cheek, as an example of how *Tass* had become more like a tabloid since *glasnost*. Some Western reporters also noted that the Russian people are far more accepting of psychic phenomena than Americans.[27]

The system has dealt with reports about extraterrestrial intelligence for quite a long time now, primarily in the form of UFOs. So far, all the characteristics described above have been displayed—

interest in anything really different, continuing skepticism by the mainstream press, sensational acceptance by other organizations.

The scientific community's near-unanimous skepticism, echoed by government authorities, has been critical to the treatment of UFOs. That response has tended to restrain judgment in spite of widespread public belief that UFOs are a form of contact. How, then, would an announcement of contact be received, assuming that it would be made by eminent scientists funded by a government agency or a credible private source?

Announcing Contact

People constantly ask SETI scientists how the public will respond to an announcement of contact. The scientists do their best to provide an opinion, even though sociology is not their field of expertise. Dr. Bernard Oliver of the NASA SETI project says:

> My own feeling is that there would be a flurry of excitement, and then when it became clear that additional information would be very slow in coming, TV sets would begin to be returned to the National Football League, or whatever. It is a parallel to *Apollo,* as far as I'm concerned. I'm not saying that bitterly, it's just human nature.[28]

Dr. Oliver's answer is representative of what many SETI researchers believe, it is close to what many journalists think, and it is probably generally correct. Similar expectations were voiced at Harvard University in 1986 when a group of SETI scientists gathered to mark the inauguration of the Planetary Society's Project META.[29]

However, it is not a definitive answer regarding impact. The initial response is going to depend on the extent to which people are prepared for the news, the nature of the announcement, and the extent to which the protocols now being established are followed. There certainly will be a confusing period during which the validity of the announcement is determined, followed by a time in which implications will be discussed.

The issue may well recede from the front pages, but that does not mean that the social impact will actually diminish. The meaning of *Apollo* is still being discussed and debated some two decades after the landing, and there is good evidence that the landing's impact is growing over time, rather than diminishing.[30]

Understanding impact requires a much closer look at what is likely to happen, based on what is now known. First, there is the issue of the extent to which the scientific search for extraterrestrial intelligence is being covered. Overall, it is not gaining a lot of attention.

As David Baron puts it, "It's the kind of thing that makes a good feature now and then; it is certainly not front page news."[31]

Other journalists confirm this view of the current level of media concern about SETI. However, interest in the topic is growing, especially as the NASA project moves closer to becoming operational, and generates more publicity.

A study by Dr. Donald Tarter of the Department of Sociology at the University of Alabama in Huntsville, surveyed a broad sample of science journalists and members of the SETI community regarding a wide range of SETI-related matters, including the handling of a contact announcement.

Dr. Tarter sent questionnaires to 221 science media people and sixty-five members of the SETI community. Fifty-three journalists (24%) responded, and thirty-six SETI researchers (55%) responded. He presented initial results of his work in a paper delivered at the fortieth congress of the International Astronautical Federation (IAF) in Spain in October, 1989.

Dr. Tarter found that both the journalists and the SETI community considered SETI to be extremely important. Ranking announcement of contact on a one-to-ten scale in terms of its importance in the history of science, the SETI group ranked it as a 9.00, and the media as 8.28.[32]

In regard to dealing with contact announcements, both groups expressed deep concerns about false alarms, but the media appeared to be quite trusting toward any announcements made by the scientists. On a 10-point scale, with 10 indicating an extremely cautious response to an unconfirmed announcement, the media gave themselves a 4.5 score.[33] That score is in contrast to expectations expressed by journalists interviewed for this book, who felt that the media would be very cautious about an announcement.

According to Tarter, the difference may be related to the fact that all of the journalists inverviewed for this book were American, whereas his sample included people from many different countries. In some of those countries, the skeptical tradition of our press may not be quite so strong.[34]

Members of the media voiced no objections to the use of announcement protocols and contact verification committees, but they did express doubt that these mechanisms would work. 57% felt that it would be possible to circumvent the work of the verification committee, and 74% expected the protocol agreement to be violated.[35]

It's worth noting that the contact verification committee, as envisioned originally by Tarter, is a body that might be appointed prior to an announcement of contact, and could play an educational role long before contact might occur.

It might be a good idea to establish the committee early, not only because it can educate the media and others, but also for committee members to begin working together before they are thrown into a highly stressful announcement situation.

Both the NASA and international protocols require an extended confirmation process following detection of a signal strongly suspected of being ETI. During this time, other sites would be asked to confirm the signal.

The initial confirmation period is critical to the unfolding of the announcement as currently envisioned by many SETI researchers. However, it will be extremely difficult to keep something so momentous a secret, and many people will have to know about it, in order to make the tests for confirmation.

Even if the news can be kept quiet during that period, the NASA protocol itself calls for a notification of NASA Headquarters, which in turn would notify the executive branch and Congress. At that time, the expert committee is convened to determine the nature of the signal and a news release is issued that announces acquisition of the signal, but not its source.

Most of the journalists interviewed for this book doubted that the protocol process would hold up for very long. Leonard David, editor of *Ad Astra*, the magazine of the National Space Society, says:

A lot will be bungled when it comes to confirmation of a signal, especially when the press becomes involved. . . . If it takes one to two months to verify a signal, the story will leak and it will break. . . .[36]

If it turns out that the signal is eventually verified, the breakdown in the announcement process will not be too significant a problem. However, if it proves to be a false alarm, that could harm the credibility of the search for a long time.

If a press release is issued suggesting that an unknown signal has been received and a group of experts is meeting to determine whether it is a product of extraterrestrial intelligence, the press will become extremely agitated. Their conflicting needs to be on top of the story while avoiding mistakes will make the announcement difficult to handle. Tension will build, as the media pressures the scientists to come up with an answer, which may be hard to do.

In addition, it is by no means certain that an announcement of contact will come from researchers who subscribe to the protocol. Someone could call a press conference and reveal their findings without submitting to the rigors of confirmation or expert committee verification. That would leave other researchers uncertain as to how they should interpret the announcement, which in turn would create confusion in the media.[37]

When questioned about these scenarios, some journalists think that any initial announcement would be met with skepticism.

According to Hugh Downs of ABC News:

A healthy skepticism should greet any news bulletin of that kind, because when quasars were first discovered . . . we said, "Wait a minute, here is something unnatural." Well, it turned out to be a neutron star spinning unnaturally fast because of the tremendous gravity. But it sure did seem like a frequency of some kind, a carrier wave or something.[38]

David Baron agrees that there will be skepticism by the media because the scientists will be unable to fully commit themselves to the idea that the signal is of intelligent extraterrestrial origin:

I think anyone would rightly be highly skeptical. We are not talking about a Martian landing on Earth. . . . We are talking about complex arguments based on signals coming from far away, so I'm sure even NASA scientists would couch it in language saying, "They really don't yet know. . . ."[39]

Ted Bonnitt agrees that if contact comes in the form of a radio signal, that will make it more difficult for people to accept its reality.

He distinguishes contact as being "organic," a landing by extrater-restrials, vs. "electronic," the receiving of a signal. He suggests that most of us have been conditioned to think about the former, but not the latter.[40]

During the initial announcement phase of this process, the critical variable is the nature of the signal. If the signal is unmistakably ETI, but yields no information comprehensible by humans, doubts and skepticism will remain. However, if there were recognizable infor-mation available as part of the signal, or (in David Baron's words) if the extraterrestrials begin sending "Maxwell's Equations," then there would be a shift in how the story would be treated.[41]

John Noble Wilford, Science Writer for the *New York Times*, covered the *Apollo 11* moon landing, as well as many other science stories for the *Times*. He said that he had once been asked what story might overshadow *Apollo*, and he spontaneously replied, "If we ever found intelligent life beyond Earth, especially beyond the solar system."[42]

According to Wilford, the discovery of intelligent extraterrestrial life would transcend *Apollo* "by a country mile," and would be "the biggest science story of all time."[43]

He believes that the interest in the story would not die down as rapidly as some might expect. Most science stories unfold over a long period of time while the experiments go through peer review and replication, he said. In the case of the *Apollo* landing, the astronauts went, they landed, they came home, and there wasn't much to write about after their return.

According to Wilford, the SETI story will not be fully developed when it is announced. Rather, the announcement is just the begin-ning, and it will have a magnetic pull on the media, as they wait for new developments. "Just because it doesn't stay on page one every day doesn't mean it won't be having an impact," he says.[44]

How would a major metropolitan newspaper like the *Times* cover the story? According to Wilford, the paper would focus on key questions, such as "Can we decode the message?" and "How should we respond to it?" and "Can we get back in contact with them?" In addition, the more philosophical issues would be addressed, looking at implications for the future: "What does it mean to find out we are not alone?" and "What does it mean to find out we are not the most advanced species?"[45]

In Wilford's view, the story would have staying power because every new announcement would have implications, and the paper would pursue those implications as well as the facts that emerged from the story.[46]

Kelly Beatty echoes Andrew Chaikin's belief that astronomers, especially the SETI project scientists, would become overnight "superstars" of the media:

> An analogous situation might be with the *Apollo* astronauts, who suddenly were catapulted . . . to the twelve most "in-demand" people on the face of the Earth. Everybody wanted to touch an astronaut, or talk to an astronaut, or meet an astronaut.[47]

These comments represent a more complete view of media response than simply that there will be an upsurge of interest after which the story will disappear from public awareness.

Society must go through an absorption period in which its concerns with an issue appear to be at a low ebb, especially if that issue challenges existing belief systems. The implications of the event are processed at deep levels of the collective subconscious, and may burst forth again into the public eye at some time in the future.

In discussing this issue with Lynn Harper, formerly of NASA's SETI Project, we agreed that there might be a "species level absorption requirement," and that this phenomenon might have something like a twenty-year cycle. There might be a period of great excitement, trying to fit new information into the old belief systems, and then a quiet time while society tries to build new paradigms that are needed in order to absorb this knowledge.[48]

The Audiences

The impact of contact will first be felt by the small group of scientists who are involved in SETI work, then by a larger group, the media. Once the message is validated and the media reports it, the entire planet will begin to feel the impact of it.

The media is only one component of the worldwide information processing system. The viewers, readers, and listeners, or "public," are the recipients of the information, and they are the reason for the exercise. They are increasingly active seekers of information, both influencing and being influenced by those who create and disseminate the news.

We often ask the question, "How will the public respond to acquisition of an intelligent extraterrestrial signal?"

This question doesn't mean very much when it is stated so broadly. Just as there is no monolithic "media," with a single approach to handling the news, there also is no "general public." The first axiom of public relations is to realize that there are many publics. Each has its own agenda, set of beliefs, and ways of dealing with new information. Messages are not "value-neutral"; they mean many things to many people.

The announcement of contact will carry with it a far different meaning to the National Security Advisor for the President of the United States than to a middle manager in a California computer software company. It will have another meaning to the director of a large Japanese trading company and a university student in China.

Predicting responses is complex, but there are some established communications research guidelines that help to fill out the model. "Diffusion of innovation" theory is an especially useful body of knowledge, based on studies of how societies adopt new products, ideas, or values.[49]

According to this research, people can be divided into five groups in terms of how they respond to innovation. These groups, and their percentages of the total population, are: Innovators (2.5%); Early Adopters (13.5%); Early Majority (34%); Later Majority (34%); Late Adopters (16%).

Innovators and early adopters see change as being in the interest of themselves and society, while the other groups are less certain about it or are actively opposed to change. Contact with extraterrestrial intelligence is an innovation that some groups and individuals will use to their advantage, while others will see it as a threat. Those already involved with SETI are clearly innovators who consider a successful search to be a positive development.

The spectrum of response from innovators to late adopters demonstrates why there are upsurges of interest in something new, followed by an apparent diminution of activity: The wave of adoption moves through the society from the most receptive to the least receptive.

Humanity's reaction to contact will be composed of a mosaic of reactions by different individuals, groups, institutions and nations. However, that does not mean that impact cannot be usefully studied in advance.

Dr. Mary Connors has examined the potential impact of contact for NASA. She notes that the impact issue is unlike other components of the SETI question, because it concerns something about which we already have a body of knowledge—human society. Moreover, where knowledge does not exist, it can be gathered through further research.[50]

Dr. Connors proposes that additional research be conducted into the attitudes of specific publics about extraterrestrial intelligence, and that such research ought to be an integral part of the SETI effort. In the meantime, we can use existing data to project possibilities. Just as scenarios can be developed to describe how the contact might occur, baseline belief systems can be established, for example, to help understand how the message of contact will be processed.

At the greatest extreme, some percentage of the world's population will simply never hear about the contact report. That seems hard to believe in a "global village" interconnected by computers, satellites, and every form of communication technology imaginable. The truth is that while everyone lives in the global village, not everyone is plugged into it.

For example, a recent poll asked Americans whether the earth revolved around the sun or the sun revolved around the earth. 21% believed that the sun revolves around the Earth and 7% did not know the answer.[51] The Copernican model of the solar system has existed for four hundred years, and it is the fourth mindstep of humanity, according to Gerald Hawkins. Still, millions of people missed that step—they don't have that "meme" in their brains.

If our analysis of how the media will cover the story is accurate, billions of people *will* hear about it. Of those who do get the news, some will believe it and some will remain skeptical.

According to Dr. Connors:

The single most important determinant of whether an individual will accept or reject the message of detection is the strength of his belief concerning the existence of extraterrestrials prior to detection.[52]

There will be those who will not believe *any* announcements about the discovery of extraterrestrial intelligence, just as there are people who do not believe that human beings went to the moon.

According to Connors, existing research suggests that about half of the population believes that extraterrestrials "have already visited this planet."[53] Many of the believers will consider the contact report unimportant because they think that far more dramatic events have already taken place. They will have no problem integrating a report of extraterrestrial contact into their belief system; for some involved in UFO investigations, there is a deeply-held belief that "the government knows all about UFOs, but is covering it up." They may argue that the government has finally realized it cannot keep the UFO story under cover any longer, and are using SETI to confirm the existence of extraterrestrials, but placing them far away, interacting with us in a manner that does not disturb the status quo.

Dr. Connors' comment about beliefs is directed at the issue of whether a report of contact will be believed or not. Her point can be generalized to support the idea that all individuals and groups will interpret the announcement of contact through the filters of their existing belief systems.

Religious views may be the most important barriers to acceptance for some. Fundamentalists of all faiths may be suspicious of reports about extraterrestrials if the reports do not conform to what that particular religion's beliefs entail.

Other faiths, such as Buddhism, are less likely to have problems with accepting a contact announcement. When *Apollo 11* landed on the moon, the Dalai Lama, leader of a major portion of the Tibetan Buddhist population, said:

We Buddhists have always held that firm conviction that there exists life and civilizations on other planets in the many systems of the universe and some of them are so highly developed that they are superior to our own. . . .[54]

Contact with extraterrestrial beings would fit easily into the belief system of the Dalai Lama's followers, and they would have positive expectations of the results, given their spiritual leader's further comments:

> We can visualize earth people journeying to faraway planets, and opening up communications and relations with beings out in space. Man's limited knowledge will acquire a new dimension of infinite scope, development and dynamism. In this, the high degree of civilization developed in other planetary bodies will be of colossal help.[55]

The Dalai Lama's expectation about communication with extraterrestrials is similar to that of Carl Sagan and other SETI scientists, and puts Tibetan Buddhists solidly in that percentage of the population who would not only believe, but welcome, a contact announcement.

There will also be national and cultural variations on the theme. For those countries with active space programs, the preparation for contact may be greater. The source of the announcement will make a difference, as will the behavior of national groups monitoring the signals. How would Americans respond to the announcement coming from the Soviet Union; will all the information be freely shared after the initial breaking of the news?

In the long term, religion may be a positive force in helping people integrate the new information comfortably. In a recent issue of *Ad Astra* magazine, experts on world religion agreed that most faiths would not have a problem with the proven existence of extraterrestrials. According to Ted Peters of the Pacific Lutheran Seminary in Berkeley, California, who has studied the possible reactions of different religions to extraterrestrials: "I believe that the majority of the religions of the world would simply accept the existence of extraterrestrial life."[56] We might imagine that some religions would consider extraterrestrials as an opportunity for conversion to terrestrial faiths.

Governments can also play a more or less positive role depending on how well prepared they might be, and the extent to which policies are in place for responding to an announcement. However, governments, like individuals, focus on issues demanding immediate attention, which means that relatively little forward planning is being done in terms of policy development right now.

John Logsdon, Director of the Space Policy Institute at George Washington University, conducted a study of SETI that attempted to find a policy analogue. He and his co-author compared SETI policy to the reporting of a nuclear accident and predictions of earthquakes. Preparation for such events reduces negative impacts, but prediction is difficult and itself has impact.[57] According to Dr. Logsdon, there is relatively little being done on SETI in the space policy field because there are so many other issues demanding immediate attention.[58]

Logsdon suggests that the international protocol effort will be the vehicle that moves SETI higher on the public policy agenda.[59]

The governments and international institutions of Earth, beset by other challenges, simply are not yet ready for SETI. If contact occurred in the near future, the response of the government/policy-making audience would very likely be uncertain at first, taking time to coalesce.

In this book, we can only begin to understand how the media and key audiences will respond to a successful contact. Social scientists should make further research on the reaction of different components of the worldwide information processing system a high priority.

Immediate Impact: Summary

The conventional wisdom on impact is that there will be a flurry of initial excitement, followed by a period in which the media moves on to other things, and the public goes back to whatever it was doing before. The conventional wisdom may be correct in how the reaction will appear on the surface. However, a far more complex response will also be taking place.

One problem in understanding impact is that the available language seems inadequate to comprehend an unprecedented event of this nature. Earlier, we used the metaphor of a worldwide information processing system to evaluate how the new information regarding extraterrestrial intelligence would be handled.

The image of Earth as a "superorganism" composed of a natural system, human system, and technosystem complements that of the information processing system. This metaphor is implicit in the

"Gaia Hypothesis," formulated by Lovelock and Margulis, the "Global Brain" concept of Russell and explicit in the idea of the "Overview System," described in *The Overview Effect*.[60]

Contact is a recognition by this superorganism of another member of its "species," and if identity is shaped by recognition of difference, it is also formed by seeing oneself in others.

Contact suddenly confirms that planetary identity by producing a mirror image of itself. What occurs next may appear chaotic and unpredictable on the surface, especially if we only observe the parts of the organism and their behavior. That would be like only watching a person's arm to understand the act of writing, or analyzing the first few weeks of the Native American response to the arrival of Europeans to understand overall impact.

What happens after the initial shock of contact is what happens when any two organisms meet one another—each one asks, "Is this going to contribute to my survival or not?" The fact of contact is highly significant, but it is only the beginning of a long-term process that will be necessary to determine long-term impact, as the two cultures interact with one another.

Chapter Nine

Long-Term Impact

*Scenarios of the impact on human society of radio
contact with an extraterrestrial civilization vary widely
between paranoid projections that contact with advanced
extraterrestrials would quickly devastate the human
spirit, and pronoid predictions that the extraterrestrials
would be so advanced that they would swiftly and
benevolently lead us into a golden age.*

—Ben Finney, "The Impact of
Contact"

THE COMPONENTS OF THE BELIEF SYSTEMS THROUGH WHICH PEOPLE
will view contact are assumptions, ideas about reality that are
stated in the form of "if/then." Thus, we say, "If we are contacted by
extraterrestrials, then . . ."

Taken together, these assumptions form a mental model of the
impact of contact which has not yet been made explicit. However,
this model should be analyzed now, before contact is made. The
assumption of the Aztec emperor, Montezuma, that Cortés and his
men were gods proved to be unfounded and fatal to Montezuma and
his people.

Behavior follows from assumptions, and human behavior in response to contact may be critical to the survival of the species. Let's
examine current assumptions about the impact of contact, and then
build a new and better model for perceiving and preparing for it.

The current model of impact includes several unstated assumptions.

The key assumptions include the following:

1. **The assumption of extreme results**—that the impact of contact will be extremely positive or negative
2. **The assumption of clear intent**—that the intentions of the extraterrestrials are a guide to understanding impact
3. **The assumption of automatic unity**—that humanity will immediately unite in the face of the "others"
4. **The assumption of one-way impact**—that the primary impact of contact will be felt by humanity

The Assumption of Extreme Results

Ben Finney has succinctly stated the two extremes in the speculation surrounding the impact issue, and has coined a term, "pronoid," to describe the optimists on this subject. In contrast to the paranoid person, the pronoid assumes that "everyone around them has their welfare at heart."[1]

Finney also says:

> Judging from the record of cultural misunderstanding between closely related human groups, comprehending a totally different civilization light years away, and absorbing the meaning of whatever messages were sent, would be a slow and tedious process calling for the efforts of specialists from many disciplines as well as the SETI scientists now engaged in the search.[2]

According to the assumption of extreme results, contact will either be the most wonderful thing that ever happened to humanity, or the most terrible. To the anthropologist, life is not that simple, and Finney's view is echoed by others in his field. Jim Funaro, who founded the annual *Contact* conferences, believes that avoiding major misunderstandings will be critical to structuring any long-term relationship with extraterrestrials, and he urges that we not prejudge the outcome of first encounters too rigidly.

Noting the numerous misunderstandings between the Human Teams and Alien Teams that have occurred in contact simulations, as well as in real intercultural and interspecific contacts on Earth, Funaro says:

> You simply cannot assume that your behavior means the same thing in another culture. Anthropologists have become convinced that peo-

ple who grow up in different cultures really perceive different realities . . .

I am not arguing that there is no objective reality, but that the reality we believe in is built for us by our culture, largely through language, from the time we're old enough to learn. People naturally interpret their own cultural consensus as objective varification and judge different views of reality as being wrong.[3]

Human beings tend to think that the objective reality "out there," is the one they are perceiving, and that everyone can perceive it that way, if they only would. However, Funaro says that this idea does not match observed responses, either with species or with cultures. Species experience different realities because they have different sense receptors. People experience different realities because they have contrasting cultures.

Communication with extraterrestrials will require far more than simply picking out their signals. On the interspecies level, pessimists point out that "We've lived with intelligent life (whales, dolphins, chimps and so on) right here on Earth for years . . . and we've never been able to even get a clue as to what they're saying."[4] As another scientist put it, "Try to talk to an ant. Now try to imagine that it's up to the *ant* to make contact with *you*."[5]

Funaro is more positive about our ability to communicate with terrestrial intelligence, such as chimpanzees, but sees a barrier with extraterrestrials if the communication is not face-to-face, because we cannot mimic behavior:

The real problem in SETI is not being able to observe them in action, because all the animals we know about on this planet have evolved a system of communication within their own species. Unless you can observe that and try to mimic it in communicating with them, you are at a real disadvantage. . . . In primate behavior, when I'm interacting with a monkey or chimpanzee, I know their behavior, so I use it, and that puts us into communication.[6]

At the cultural level, the number of misunderstandings and problems that have arisen in a multitude of encounters on Earth are too numerous to catalog here. However, a few examples will show that misunderstanding is probably the rule rather than the exception.

Finney points to the contacts between tribespeople in New

Guinea, living at a Stone Age level at the turn of the century, and the representatives of Western colonialism who discovered them. A misunderstanding based on different perceptions of reality led the New Guineans to construct a new belief system, known as the "cargo cults," that grew out of their old beliefs coupled with the contact experience.

After initial contact had occurred, the Westerners tried to get the New Guineans to work on their plantations and become involved in the planting of cash crops.

However, many of the New Guineans who began to work the plantations abandoned it and tried a different approach:

> Villagers living on the coast would build a crude wharf, and then erect a "radio station"—made of wood, bamboo, and vines. They then would spend days dancing, carrying out elaborate rituals, gathering at the "radio station" to send messages, and waiting on the wharf for a ship to come.[7]

Farther inland, where bi-planes brought the "cargo," villagers built crude airstrips and tried to attract airplanes to land there. They believed the way to get the material goods of the Westerners was to conduct these rituals that "attracted" ships and planes with cargo.

The Western officials tried to convince the native peoples that the goods they wanted could only be purchased with money earned by work, not by rituals. The cargo cult adherents refused to believe them, because they had already seen how little the money they earned on the plantations or for growing their own crops had bought. They believed that the officials were lying to them in order to keep their own secret about "cargo."[8]

Today, the United States includes many different cultures that began their relationships based on misunderstandings that continue to this day. Native Americans and the descendants of European immigrants, for example, are still struggling to communicate across a great cultural divide. African-Americans and whites have the same problem—and what is racism but an inability to comprehend those differences without fear and prejudice?

Similar conflicts exist on an international level. The United States and Soviet Union allowed their differences to threaten the very existence of the species for some forty years, before a transformation in the relationship began in the 1980s.[9] The Arabs and Israelis are still struggling to find common ground, and many other regions of the world are far from unified or cooperative.

Is one side totally right and the other utterly wrong? If Jim Funaro is correct, it is more accurate to say that they are both operating honestly within their own cultural realities.

What does this terrestrial history mean for the assumption that contact will have extreme results, either good or bad? It suggests that the results will actually be mixed, with some positive benefits and some truly negative ones, and with the effort to understand one another being of the highest priority.

The Assumption of Clear Intent

It seems logical to ask the question, "Well, what are the intentions of these extraterrestrials?" If the extraterrestrials are benign, then the outcome of contact should be good, shouldn't it? And if they are hostile, then the outcome will be bad, won't it?

Closer examination shows that determining intent is difficult in itself, because of everything that has been said about cultural mis-understandings. Moreover, results do not directly follow from inten-tions in the complex universe of social evolution. Even if we can determine the intentions of the extraterrestrials, it's possible that benign aliens will create real problems for humans, and hostile aliens might be a blessing in disguise.

In addition, it is not clear that we totally understand our own intentions in searching for extraterrestrials, so how can we be so confident of grasping theirs?

It may not be easy to predetermine the intentions of the extrater-restrials, but it is possible to illuminate current human assumptions about those intentions. Here, a remarkable evolution has occurred, from a view of extraterrestrials as a threat to humanity, which prevailed some forty years ago, to a consensus that they are probably benign.

The popular culture has exerted a powerful influence on our conception of extraterrestrials and aliens. In the 1950s, a time of great insecurity and paranoia on Earth, most aliens were seen as hostile, a threat to the planet that had to be destroyed or repelled. This attitude built on the earlier "War of the Worlds" vision of Martian invaders initiated by H. G. Wells and popularized by Orson Welles in his 1938 radio broadcast.

By the 1980s, a more confident view of extraterrestrials, personified in E.T., Close Encounters of the Third Kind, and Starman, began to emerge. In these films, the hostile forces are represented by government and military officials who fail to understand the extraterrestrials and want to dissect or destroy them. On television, "Alf" (Alien Life Form) manages to fit right into a suburban American family, functioning like a guest who came late to the party and never leaves.

The old concerns have not entirely disappeared, however. The two films, Alien and Aliens, present a view of an implacable enemy of humankind, a menace that can be stopped only by fighting back with an equal fury. In addition, an updated "War of the Worlds" has been a continuing television series. The Martians have returned, still intent upon taking over the earth, and escalating their strategy by taking over human bodies. This is reminiscent of The Invasion of the Body Snatchers, a truly frightening film because the principal character is forced to mistrust humans as well as aliens.

Alien Nation, another film that became a television series, presents an even more sophisticated version of the culture-conflict problem. In this treatment, the aliens do not come to Earth to conquer it, but they do set up their own nation within a nation here. Unfortunately, they also bring with them a terrible secret about their own culture, which is threatening to human society.

All of these scenarios are a subconscious effort to deal with our normal and natural hopes and fears about the nature of the universe. It's understandable that fear is a strong component in some of these simulations of the future. Humans have been forced to struggle against overwhelming forces in order to occupy the evolutionary niche that they enjoy today. As Hugh Downs puts it:

Look at our pre-history. Here we are, an unspecialized animal. We didn't have speed, or bulk, or fangs or claws, but we did have a brain. What an existence we must have had, back in prehistoric times when life was nasty, brutish and short. We were easily threatened by every kind of disaster that came along. . . . Think how brave humans are. . . .[10]

This process of preparing human beings for what may come in the future is a healthy one, and mirrors the "paranoid/pronoid" distinction Ben Finney has described. Star Wars has been analyzed by Joseph Campbell and shown to have the basic structure of all great myths: the hero undertaking a dangerous journey, encountering obstacles and strange creatures, then finding wisdom in a process of inner self-discovery. The extraterrestrials in Star Wars are like human beings in real life: They are both good and bad. Some (like "Yoda") are incredibly good, and others (like "Jabba the Hutt") are unbelievably evil.

Yoda and Jabba are different, but still comprehensible to humans. Yoda is an archetypal Zen master, while Jabba reminds us of a Mafia boss. The message of Star Wars is the same as the message of most such stories. Regardless of the scale or locale of the conflict, good wins out over evil.

Beyond the myths and assumptions, is there a way to determine what we can expect from the universe? Let's begin by examining the assumption that any civilization contacted by SETI will be benign because "they are more advanced than us."

This assumption is built on yet another assumption, which is that it's appropriate to use the term "advanced" in this context. Certainly, such a civilization is likely to be older than ours, and it must be at least on a par with us to have the technology for interstellar communication.

Michael Michaud has consistently suggested that we must think far more broadly on this issue. At one point, he wrote "an alarmist scenario" to counter the model being disseminated by Carl Sagan, which, he says, focuses on "a high-minded exchange by radio signals of scientific information."[11]

We want to assume that any extraterrestrial civilizations contacting us have moved beyond our own violent and confusing historical period, and have resolved many of the problems that plague us. However, as Michaud points out, there are many possible scenarios of advanced evolution on planets circling distant star systems.

Today's human society is one thousand years older than the

human civilization of Europe in 990 AD. Society has evolved dramatically in the past thousand years, but many of the conditions
that existed in 990 persist today. There was poverty, war and disease
then, and there are poverty, war and disease now. Interspecies
contact has accelerated during that period, often resulting in extinction for the non-human species. A multitude of "culture contacts"
have occurred during that period, and many cultures have been
traumatized by them. Many elements of human life have improved
in one thousand years. However, it is a legitimate question as to
whether Earth-based societies are much more benign today than in
the past. Will they be any more so in 2990?

We have no knowledge base available to us about physical and
biological evolution on other planets, except within our own solar
system. How, then, can anything be said about social evolution?
Given the diversity of cultures on Earth, it makes sense to assume
that civilizations may take different and unpredictable forms within
our galaxy and beyond.

Isn't it possible to imagine a culture with the technology to
communicate and even travel in space, but with the mindset of a
latter-day Roman Empire? Rome was an advanced civilization for
the time, and it was also imperialistic, dominating "less advanced"
cultures, such as the Gauls. Rome made tremendous contributions
to the world in the areas of law, administration, and engineering,
but it was also a society built on the backs of slaves. The Roman
example pinpoints the problem in defining "hostile" and "benign."
Rome saw its rulership as a benign gift of civilization to backward
peoples, while the subject peoples no doubt saw it as being extremely hostile.

When the North American continent was being settled, the settlers
called the Native Americans "hostiles." But who was being hostile
to whom, and why? To the native peoples, the earth was their
mother, a sacred part of their lives. It was not something to be
owned, carved up, mined and exploited by others. The behavior of
the settlers, while quite normal for their culture, was hostile to the
Indians.

Some of the white people involved in that culture clash were
truly benign in their intentions, such as the missionaries. Bringing
Christianity to the people of North America represented the most

precious gift they could imagine. However, their efforts were a hostile act relative to the tribal culture of that time, and helped to destroy much of it.

The extraterrestrials don't have to be consciously bent on hurting Earth's society to be destructive to us. Just being themselves might be a challenge to us. As Michael Michaud has pointed out, "Human history is littered with examples of cultural shock, of cultures that have been destroyed or absorbed by superior civilizations."[12]

The 1989 *Contact* conference provides a good example of the miscommunication of intent and its consequences.

As in previous conferences, an Alien Team and Human Team spent two days evolving separate societies based on assumptions supplied to them in advance. On the final day of the conference, the two cultures met, with the rest of the conference participants looking on as an audience. The coordinator of the simulation served as a referee; no one, including the teams, knew what would happen when contact occurred.[13]

The following is a description of what happened, based on notes I took while watching the encounter:

> The humans live on a colonized planet of the Tau Ceti system, several light-years from Earth. The planet is similar to Earth, but very watery, with far less land mass. The colonists are experts in genetic engineering, and have created a sub-species known as "Tads," who are adapted to the water and work there for the land-dwellers. Some Tads rebel against their harsh fate, and return to land, where they are known as renegades. The planet is known, appropriately, as "Puddle."
>
> Puddle is supplied, as are all colony planets, by great spaceships known as "galleons," the last remnants of a once-thriving galactic (human) civilization that is now in decline. As the action begins in the simulation, the people of Puddle are waiting for the arrival of a galleon known as the *Magellan*.
>
> The *Magellan* is a multi-generational starship with thousands of crewmembers aboard. It has no home port, but spends its life cycling from planet to planet, with an average of twenty-five years transit time from port to port. The people aboard have a kind of "ship religion," based on the importance of themselves and their ship to humanity. Their value system calls for preservation of the ship above all else.
>
> The aliens are known as *Us*. Their planet is based on the world-building exercise that produced Ophelia at an earlier *Contact* event, and was described earlier in this book.

Ophelia is a hostile planet, and the life forms that became *Us* did so by banding together into a collective organism. Beginning as single-celled organisms composed of a genetic material known as XNX, *Us* evolved by absorbing other life forms into itself, creating a greater entity. The *Us* organism perceives the universe as divided into *Us* and *Not-Us*. It considers all organic life as *Us* and all non-organic life as *Not-Us*.

Anything, including heavy metals and radiation, which is *Not-Us* is considered food. Anything that is *Us* is a candidate for union and absorption into *Us*. The humans don't know this, and from a human perspective, it is difficult to distinguish being absorbed into *Us* and being eaten by *Us*.

When *Us* became too numerous to continue evolving on its planet, it created spacecraft and began to explore outer space. After a time, the spacecraft approached the Tau Ceti system and made efforts to communicate with what appeared to be other forms of *Us* on the planet Puddle and near it.[14]

When the encounter began in front of the other conference participants, a number of extraordinary events unfolded. In particular, the perception of intentions became distorted, producing destructive behavior based on misunderstanding. Everything that *Us* did to communicate with the humans was misunderstood, and the humans reacted with hostility, trying to protect themselves against what seemed to be an attack.

Us sends parts of itself down to "seed" a gas giant planet that is part of the Tau Ceti system; the humans see it as an invasion. *Us* sends its organic communications devices into the atmosphere of Puddle itself and the humans grow more concerned, seeing it as another attack. The humans send probes to try to find out what the *Us* ship might be. *Us* views the probes as food-offerings, and eats them. This creates new fears among the humans, who are now totally lacking in information about *Us*.

Us detaches a part of itself, a separate ship (*Contact Us*) from the mother ship, and puts it into orbit around Puddle. The humans send a shuttle into orbit to intercept the *Us* ship. The humans throw a huge paint bomb at the solar sail of the *Us* ship, coating the sail and making it lose power. *Us* eats the paint and enjoys it. *Us* moves close to the shuttle and tries to "slime" it, which involves putting a sticky material on it. This means "Don't eat me," in *Us* communications. The humans find this action to be even more hostile and bizarre.

Us is confused. It assumes it can eat the humans because it has never come across something organic that was not also Us. On the other hand, the humans, in some ways, act like the Us. Us would like to go aboard the shuttle and bring back a human or Tad that can be broken down into its consituent parts. Then, it would be possible to know if the being is Us or Not-Us.

The Us never get the chance. The humans become increasingly panicky, and division breaks out among the human cultures. The Magellan crew are fearful for the safety of their ship, but the Puddle people need their help in this time of crisis. The Puddle people take over the shuttle. The Magellan sends out a signal quarantining Puddle, and then moves away to a safer distance.

The humans fire lasers at the Contact Us, which the aliens try to eat. The Contact Us is injured, but not killed. The humans then fire a genetically-engineered virus into the Contact Us, and Contact Us is destroyed. The Us mother ship backs off to decide what to do next, and the simulation ends.[15]

After the encounter ended, the two teams talked about the lessons learned in the simulation. The Alien Team revealed that *nothing the humans did was considered hostile,* and the aliens had no hostile intent. The Human Team realized that everything the aliens did was meant to be benign, but was experienced as hostile. All of this misunderstanding took place without the aliens accomplishing one of their goals, which was to dismember one of the humans or Tads to find out what it was!

That would not have been seen as hostile by the Us, but imagine what the humans would have done if one of their number had been kidnapped and taken apart. The human team was composed of anthropologists, students, writers, and other thoughtful people. They probably were not a cross-section of human society, in that they were better educated, more open-minded and much more experienced in cross-cultural issues than most of us.

The simulation took on its own momentum as it developed. The team members became the people they were representing, and they reacted with concern when faced with unfamiliar actions by creatures they did not understand.

The pressure of time and the lack of information become important in simulations and in real-life situations where danger may be present. The humans in this simulation might have wanted to take awhile to learn more about the Us and make decisions based on knowledge rather than speculation. In the meantime, however, the *Us* were bearing down on them, and doing things that seemed to threaten human lives.

The humans' decision turned out to be wrong, but they were not unintelligent, based as they were on their understanding of the situation, and their feeling that something had to be done quickly. In the meantime, who can say what the impact of this short-term misunderstanding would have had on long-term relations between humans and *Us*?

To those who are interested in the results of actual contact, sometime in the future, this simulation is thought-provoking, and should cause us to question deeply our assumptions about how contact will evolve.

The *Contact* conference simulation is similar to inter-cultural interactions that have taken place on Earth when physical contact occurs, and the control of living space is involved. That is different from what we imagine happening in today's SETI process, where communication takes place without physical proximity.

In a situation where communication is happening over light-years, with long periods of time for discussion, perhaps better decisions would be made. However, if there is a paucity of accurate information, perhaps not.

The extraterrestrials might take actions that are clearly intended to help humanity, and even understood as such, but the effects could be hard to handle. Consider the scenario that many SETI advocates hope for: Suppose that a galactic civilization, composed of some 350 planets in the Milky Way, sends a beacon, and then transmits the entire *Encyclopedia Galactica* as a way of welcoming Earth to galactic society.

What could be more benign and supportive than that? Think of the immense positive changes that we could make in our society if we had medical knowledge thousands of years in advance of our own, or near-perfect weather forecasting capability. The extraterrestrials might also be quite advanced in the social sciences, providing us with insights on how to manage our societies better. Technologically, they would undoubtedly have a number of marvels to share, especially with the use of the *Encyclopedia*, drawing on the knowledge base of the entire galaxy.

For some, this would be a marvelous opportunity for learning, as new vistas of knowledge opened up to them. But what effect will it have on the human psyche as a whole to discover that no matter how much we learn, that knowledge and much more is already known?

Suppose you're a thirty-five-year-old molecular biologist at a major university. Your work in genetic engineering has won many honors; there's talk you'll win the Nobel Prize next year. Then, the day of contact comes: The *Encyclopedia Galactica* is transmitted and the first volume contains genetic engineering information so far beyond you that it will take years to comprehend it. Far from being the leader in your field, you must become a student once again. Will you feel exalted or depressed? Will a new mental illness result from contact that resembles manic depression on a grand scale?

I discussed this issue with Dr. Bruce Shackleton, a psychologist with appointments at Harvard Medical School and Massachusetts General Hospital in Boston.

According to Dr. Shackleton, the impact of contact on individuals will depend upon the extent to which they are focused on themselves and their own egos as opposed to being open to wider horizons. If the researcher in question is driven by an honest pursuit of knowledge, he will be delighted at the advent of the extraterrestrial information. If he is driven by the status and prestige derived from his work, he will be disoriented by it.[16]

Dr. Larry Kaye, a Boston-area sociologist, confirmed this in terms of social impact. He observed that much of our thinking is very egocentric, and our paradigms are organized around humanity being the primary species on the planet. If we contact a species that is clearly more advanced than we are, it would force a tremendous rearranging of our worldviews.[17]

Impact will extend to every other subsystem of society, including the economic and political sectors. Consider, for example, the impact that the *Encyclopedia* would exert on economic systems that thrive on innovation and unpredictability. What would happen to the stock market if all of IBM's computer technology became obsolete overnight, or all of the automobiles of the world were suddenly as archaic as a horse and buggy? Suppose that the extraterrestrials know how to travel faster than the speed of light, and come to visit us.

The existence of a galactic society, even if it is a peaceful, benign civilization, means that there will be social conventions to be obeyed by newcomers. The underlying issue is whether Earth's society will become stagnant if humans find that we are not only not alone in the universe, but one of the more recent arrivals in a crowded and well-developed galactic civilization.

The gathering of new information through physical and mental exploration has been a driving force in social evolution over the millennia. As new knowledge appears, it breaks down old molds, and prevents social systems from rigidifying. If people feel that everything is already known, what will they do in response to that feeling, and what will happen to the mechanism of social evolution?

Science fiction author David Brin writes about this dilemma in his novel, *Startide Rising*. He depicts a galaxy filled with a variety of species, many of them implacably hostile to humans. He imagines a giant Galactic Library in which all knowledge is stored. Most of the civilizations of the galaxy use the library without questioning it and are developmentally frozen because they have lost interest in innovation. New species that enter into the galactic society must be "uplifted" and serve as clients to patron species for as long as one thousand years.

Humans throw the process into disarray by refusing to serve as a client species, claiming that their genetic engineering of dolphins and chimpanzees has brought sentience to those species, qualifying humanity to be considered patrons. Humans also try to abstain from using the Galactic Library because of its debilitating effects.[18]

Is Brin's vision of the galaxy prescient? Might it be more accurate than the expectation that contact will lead to a "high-minded exchange of scientific information?" The answers to these questions are unknown for now, but the assumption of extraterrestrials with benign intent is clearly questionable. Moreover, the effects of the extraterrestrials on Earth may have more to do with our response than their intentions.

Let's also consider the other side of the coin. Suppose the extraterrestrials are hostile to humanity. Is it possible that bad intentions on the part of the extraterrestrials might produce good results for humans? At this stage in human evolution, our species may be better prepared to respond to a hostile challenge than adjust to no longer occupying the top niche on the evolutionary ladder.

It has been remarked on many occasions that such an eventuality might be a road to world peace, and this is the foundation of another important assumption, the assumption of unity. However, even this assumption cannot remain unchallenged once it has been examined carefully.

The Assumption of Automatic Unity

SETI does create pressures toward planetary unity and international cooperation through activities like the protocol movement, which seeks to generate a planetary approach to disseminating news of contact, and to determining Earth's response. Dr. Michael Papagiannis points out that early in the history of bioastronomy, participants concluded that theirs was an effort of all Earth, not of individual nations. Members of this field have created their own flag as a manifestation of their sentiments.[19]

There is a widespread belief that if Earth were threatened from outer space, all humans would unite to repel the invaders. Former President Reagan alluded to it on several occasions. Once, in discussing the most important issues in international relations, he told an audience:

> I've often wondered what if all of us in the world discovered that we were threatened by a power from outer space—from another planet. Wouldn't we all of a sudden find that we didn't have any differences between us at all, we were all human beings, citizens of the world, and wouldn't we come together to fight that particular threat?[20]

Contact would force humans to perceive a new level of one-ness among themselves in contrast to beings from another planet or galaxy. The entire concept of "aliens" brings forth a sense of strangeness, of "other." However, unity is not automatic; it must still be created.

Even now, SETI groups are not united on how funds ought to be spent on the search, nor whether there should be a protocol for responding to a signal. Examples from other cultural contacts suggest that the extent to which competing human groups create unity in the face of external threats depends on prior relationships as well as the perceived threat.

For example, unity was not the result of the European incursions into North America. The native American tribes at the time had a tradition of allying with the strongest available tribes against their enemies.

Many tribes, such as the Apache and Sioux, fought ferociously against the white invaders, and only surrendered when they were beaten by overwhelming odds. Others, such as the Crow and Blackfeet, allied with the whites against their traditional enemies, including the Apache and Sioux. Some chiefs saw that the whites would be the dominant force in North America, and cast their lot with the new arrivals. They even punished their young braves if they took up arms against the pioneers and wagon trains moving west. The native Americans never overcame their natural differences sufficiently to respond to the whites, and eventually they were defeated.[21]

We should note that the strength of the native American cultures is such that they have survived that early defeat, maintaining their traditions through extremely difficult times. There is a lesson here as well: Strong cultures do rebound from traumatic culture contacts.

The native American cultures of the mid-nineteenth century were tribal in nature, and had not evolved to the level of modern nation-states on Earth. These nation-states are already beginning to share an increasingly global and unified communications system and financial structure, and are moving toward greater cooperation even without being threatened by outsiders.

The trend toward global unity and the search for extraterrestrial intelligence appear to be complementary trends, driven at least partially by technological developments. As we reach a higher level of awareness of ourselves as part of a planetary civilization, we have begun to reach out for contact with other civilizations. From an evolutionary perspective, it may be that cultures only begin the search when they are prepared for the impact.

The native Americans were not prepared for the impact of contact. Except for the damaging myths about the "return of the gods," these societies had made little preparation for contact with another culture. By contrast, the people of Earth have the opportunity to prepare for contact. The subconscious interest in the topic is exploited on a mass basis in books, films, and television programs, even though there is not an intentional effort to "get ready for SETI."[22]

Preparation should be deeply imbedded in the SETI programs themselves, because there will be consequences of success, and it is simply not ethical to ignore those consequences. We should be considering what will happen to the world consensus if the extraterrestrials are not simply hostile or benign, but are just there, looking for some type of relationship. In that event, the world community will be neither unified nor polarized, but will begin an intense debate on what to do.

In my interview with Michael Michaud, we discussed the fact that I had initially thought gaps in technological capabilities were the key to determining how less advanced cultures performed in contact situations. It seemed that in the Americas, the indigenous cultures were overwhelmed primarily by technological superiority, but the generalization did not hold in modern times. The United States, with significant technological superiority, attempted to impose its will on Viet Nam in the 1960s, but failed. The Soviet Union made a similar effort in Afghanistan in the 1980s and failed as well.[23]

Michaud cited Japan's response to Commodore Matthew Perry's mission opening that country to the West in the mid-1800s. He said that the Japanese responded by adopting a strategy that maintained their unique cultural norms while adapting in other ways to Western culture.

Within about 150 years, the Japanese have reached parity with the Western powers in most areas, and are dominant in others. This took place even after the Japanese were subjected to a humiliating defeat in war and had two of their cities devastated by nuclear weapons. After the war, they were able to accept and utilize a Western-style democratic system imposed upon them by the victors.

Considering the Japanese example in particular, Michaud suggests that *social cohesion* might by a critical variable in determining how a technologically less advanced civilization will cope with a more advanced one.[24] The Japanese analogy shows that, in addition to social cohesion, adaptability and perseverance ought to be considered in this equation. To what extent can a society adapt to circumstances without giving up its own unique values, and how long can it persist in relationship to a dominating "other?"

Planetary unity, then, is not something that will necessarily follow from contact; it is something we ought to create in preparation for contact.

The Assumption of One-Way Impact

When culture contacts have occurred on Earth, both civilizations have changed. When the Romans conquered the Greeks, they made the Greeks into slaves, even though Greek culture was superior to the Romans in many areas.

The conquest appeared to represent a decline for Greece, and in terms of their freedom and independence, it was. However, it also uplifted the Romans and exerted a positive influence on the empire. From the Greek perspective, what happened was negative, but there were salutary effects for human civilization.

The same can be said in regard to other relationships, such as the Europeans and native Americans. While the European arrival looks like nothing but an unmitigated disaster for the Indians, this may be an overly simplistic view. The native peoples have survived and their culture continues to have impact on white civilization.

Human culture is rich and diverse, and the earth is an amazing planet. We have the ability to create a reverse impact in a contact situation, to the benefit of extraterrestrials and terrestrials.

Lynn Harper suggests that the concept of "more advanced civilizations" and "less advanced civilizations" is too narrow:

> My view is that just as every nation on this planet is more advanced in some ways and less advanced in others, the extraterrestrials will be like that. For example, there may be a species very advanced in communications, but far behind in medicine.[25]

As humanity begins to reach out to the galaxy and the universe, the real question is not "Who is more advanced?" and "Who is less advanced?" but rather, "What is the unique contribution of each society to the overall galactic culture?" We need to decide what ours will be, regardless of what we find in our explorations.

These, then, are some of the many assumptions that now form the foundation on which human civilization will build its response to successful contact. Reviewing the assumptions reveals that most of them are overly simplistic, and that we must go to a deeper level of analysis to understand-impact. We need a model of impact that takes into account many different variables.

The Contact Impact Model

At its simplest level, building this model requires, first, an understanding of how systems evolve; second, a definition of impact within that framework; and third, clear assumptions about major elements of contact.

Systems theory tells us that any system, whether it is an atom, an animal, or a human society, is constantly moving through different system states, including equilibrium, change, and transformation. The movement from one state to another is caused by new information entering into the system and having an impact on it.

Change takes place when a social system absorbs this information, but transformation occurs when the information is so profoundly different that the system's existing paradigm must be restructured to accommodate it.

The movement of systems through different states is called evolution. Evolution is produced by exploration, which feeds information into a system. If contact with extraterrestrials occurs, dramatic new knowledge will enter the worldwide information processing system, which is part of the planetary overview system now emerging on Earth. In terms of the impact on that system, it can be measured by:

> The extent to which contact affects the Earth system as measured by the degree to which terrestrial evolution is altered by the contact.
> The degree to which the course of terrestrial evolution is altered can, in turn, be understood as: the amount of new knowledge received in a given period of time.

We can be almost certain that contact will produce change in the Earth system. The question is, will it also produce a fundamental transformation?

We create Level One of the model using assumptions about two components: distance and time.

Distance: Distance has already been identified as an important factor in determining the probability of contact, and it is also critical in determining the impact of contact.

Based on all that we now know, the distance assumption is:

Impact increases as the proximity of the contacting civilization to Earth increases.

In other words, impact will increase if the contacting civilization is close to Earth, and decrease if it is far away. Specifically, if we hear from a planet ten light-years away, that will have more impact than a signal from another galaxy.

People will feel more affected if the contacting civilization is close to Earth, because there's a greater likelihood that the contact will affect the lives of individuals. The closer the civilization, the more rapid will be the dialogue, and the more likely that interstellar flight may eventually produce face-to-face contact. A greater volume of new information will be fed into the system more rapidly, causing greater changes to take place.

Time: We have already discussed the fact that the probability of contact also increases with time. We assume that impact will decrease over time, as human society matures. The assumption is, then:

Impact increases as the proximity of the time of contact to the present increases.

The sooner contact occurs, the less prepared humanity is likely to be, and the greater the impact. The farther in the future the contact occurs, the more evolved and better prepared humanity will be, and the less the overall impact.

As humanity moves out into the solar system, and space exploration becomes a greater part of our lives, more people will begin to think about extraterrestrial intelligence seriously. We ourselves will become extraterrestrials as the first children are born in space, and the first immigrants end their lives there!

Level One of the Contact Impact Model is summarized, then, as an equation which reads:

I [Impact Index] = 1/D [distance in light-years] ×
T [time in years]

The one (1) in the numerator is set as a constant that has no meaning except as a way of setting the scale for the index, and as a place-holder for values that will be inserted in developing Level Two.

In theory, the index can vary quite widely. For example, suppose Scenario 2-B from Chapter Seven (Contact with a nearby star system within the Milky Way) turned out to be the actual contact situation, as part of the targeted search being undertaken by NASA. Contact might be established with a planet circling Tau Ceti (12 light-years distant from Earth) in 1995 (5 years from the present). The Impact Index would then be:

$$Impact = \frac{1}{Distance \times Time}$$

$$= \frac{1}{12 \ light\text{-}years \times 5 \ years}$$

$$= \frac{1}{60}$$

$$= .017$$

By contrast, Scenario 2-C (Contact with a distant star system within the Milky Way) might be the actual situation, perhaps as part of the All-Sky Survey being undertaken by NASA. If the contact were with a planet circling Deneb, 1,600 light-years away, in 1995, the index would be:

$$Impact = \frac{1}{Distance \times Time}$$

$$= \frac{1}{1600 \ light\text{-}years \times 5 \ years}$$

$$= 1/8000$$

$$= .0001$$

These two examples clearly illustrate the effect of distance on impact, as contact with Tau Ceti is much more significant than with Deneb.

Level Two of the model is created by adding two other factors, *parity* and *information*.

Parity: Parity concerns the distance between the two civilizations in terms of development, with impact increasing as the gap widens.

The parity assumption is:

Impact increases as the differences in the levels of development of the two civilizations increase.

Parity is unlike distance and time in two ways. First, it isn't clear how to measure parity differences, although differences in age are probably the best approach. Second, even if we knew how to measure this quality, it isn't clear that initial contact will reveal much about it.

For example, if we make contact with a very old civilization possessing great knowledge compared to our own, we can assume that the event would have substantial impact on our society. However, a younger civilization may have evolved more rapidly than ours because of unique conditions within its solar system. Also, when first contact is made, we're unlikely to know the social development of the other culture.

Parity might also be stated in terms of whether the other society is Type I, II, or III (Kardashev's terms), or is a planetary, solar, or galactic civilization. For the sake of illustrating the model, we will simply assume a parity difference between the contacting civilizations, stated in years. This is based on the assumption that each additional year of existence is equivalent to an opportunity to create a unit of new knowledge unknown on Earth (this might be called a "Knowledge-Year").

Information: Finally, it is assumed that as more new information is transmitted by the contacting civilization, the impact on Earth's society increases.

The information assumption is therefore stated as:

Impact increases as the amount of new information transmitted increases.

This component can be seen either in terms of message content or volume of information—as "What do they say?" or "How much are they saying?" In this example, information volume is used, because it can easily be quantified, while content is more subjective.

The initial message may be very simple. It might be a beacon, like a lighthouse, sending out a continuing pulse of electromagnetic radiation. Information would be contained in the beacon because its very existence would tell us that intelligent life does (or did) exist elsewhere, and that it reached the point of being a technological civilization interested in communicating with other civilizations. The location of the beacon would also be knowable. That kind of message would have great impact, but somewhat less than a signal with more information contained in it.

The beacon can also function as the "carrier wave" for another signal that does contain information. Perhaps it would include mathematical formulae and pictures pinpointing the location of the transmitting civilization's home planet. There might be information about the civilization itself, but nothing about the rest of the galaxy or beyond. Frank Drake developed such a message as an experiment, using only 551 bits of information.

Finally, there is the possibility that the civilization would send us everything that they know, or even that an *Encyclopedia Galactica* exists, and the signal would contain that type of information. The impact of that form of transmission would clearly be enormous.

Adding parity difference and information volume to the Level One formula produces this Level Two formula:

I [Impact] = Pd [parity difference in years] × Vi [fraction of total knowledge transmitted] / D [distance in light-years] × T [time in years]

Let's take the two previous examples and calculate the Impact Index again. This time, let's assume that the transmitting civilization at Tau Ceti is about 2000 years ahead of ours and that it sends a message containing 10% of the available information. The index is then calculated as:

$$Impact = \frac{2000 \; years \times .1}{12 \; light\text{-}years \times 5 \; years}$$

$$= 200$$

$$= 3.33$$

Using the same numbers for the transmission from Deneb, the result is:

$$Impact = \frac{2000 \text{ years} \times .1}{1600 \text{ light-years} \times 5 \text{ years}}$$

$$= 200$$

$$= .025$$

Adding parity difference and information takes the model to a new level, and it provides a better impact index. However, it does not really answer the question of impact quantitatively, because it does not answer the question, "How much new information is arriving in a given period of time?"

To do that, a different version of the Level Two formula is required. In this version, it is understood that if the entire extraterrestrial knowledge base were sent and received instantaneously, the impact would be equal to the initial difference in development of the two civilizations.

However, during the time that the information is traveling to Earth, terrestrial society is also evolving, reducing the parity difference, and the "newness" of the information. This is another way of making the Level One statement that Earth society matures as time goes on, thereby reducing impact.

At this level, there is no longer a need to include time to contact. Impact is measured in terms of the difference in development experienced at the time of contact, whenever contact might take place.

This version of the Level Two formula is, then:

I (Impact) = [Pd (parity difference in years at time of transmission) − D (Distance in light years)] × Vi (fraction of extraterrestrial knowledge base transmitted per year)]

Let's take two previous examples and calculate impact again. This time, let's assume that the transmitting civilization at Tau Ceti is about 2000 years ahead of ours and that it sends a message contain-

ing 10% of all the information available to it. Impact is then calculated as:

$$Impact = (2000 \ years \ - \ 12 \ light \ years) \times .1 \ per \ year$$

$$= 1988 \times .1$$

$$= 199, \ or \ about \ 200 \ units \ of \ new \ information \ per \ year.$$

Using the same numbers for the transmission from Deneb, the
result is:

$$Impact = (2000 \ years \ - \ 1600 \ light \ years) \times .1 \ per \ year$$

$$= 400 \times .1$$

$$= 40 \ units \ of \ new \ information \ per \ year.$$

These numbers can be given more substance by defining what the
"units of new information" might be, which is discussed in more
detail in the Appendices.

The model can be linked to the Drake Equation in several ways. If
there are many civilizations, for example, they are likely to be closer
and we will probably hear from them sooner. As N increases, then,
so does I.

At its current stage of development, there are several ways to
quantify the model. As the examples have shown, actual numbers
can be used for each variable. An alternative approach is to assign
values to each component, such as a "1" for contact from a civilization in another galaxy and "3" for a civilization twenty light-years
away.

These choices are a matter of convenience and personal prefer-
ence, and the numbers themselves should not be taken as absolutes.
The best way to assign values remains an open question, and is
considered in more detail in the Appendices.

Determining how to assign values to the model is one useful area
of additional research. Thinking of additional components to be
added to the model is an even more valuable contribution; the
appendices also consider some of the components that might be
added.

Level Two: Two-Way Impact

The Contact Impact Model is valid in considering two-way impact as well, although it does require a shift in perspective to think about our impact on the extraterrestrials. The model can measure the impact of our response to a transmission from another civilization, or the impact of Earth contacting another society in the listening phase.

If the model is used to measure the impact of our contacting a society that is still in the listening phase, then there are no changes in the factors, and it is a simple reversal. "They" become "us," and impact refers to the impact on them.

If Earth is responding to a communication, the Level One elements are unchanged, with impact being proportional to their proximity to Earth and the proximity of the contact date to the present. Information volume is treated in the same way, but refers to the amount of information in our response. However, parity difference changes, because the impact on the other (more advanced) civilization will presumably decrease if the differences are great, and increase if they are small.

The model's greatest value is that it points to areas that need much more exploration. Working with it reveals many things, but one insight stands out above all others—human beings cannot control most of the factors. We cannot determine the distance of the contacting civilization in advance, nor the year in which contact takes place. The type of civilization sending the message, and the volume of information are also the prerogative of the extraterrestrials.

The one factor totally within human control is partially represented in the model, and that is *preparation*. The model implicitly includes preparation in the time variable, on the grounds that as more time goes by, humans will be better prepared for contact. However, that is a choice we may not make; the current upsurge in SETI activity may indicate a blip, rather than a trend. After all, many assumed that after *Apollo*, the United States would lead the expansion outward into the solar system and beyond, but a long fallow period followed instead.

This leaves open the question of what would be a high or a low level of preparation, a policy question that should be answered on a worldwide basis. John Logsdon's study makes a parallel with earthquake planning, which is appropriate. The impact of an earthquake is enormous no matter what is done, but preparation significantly reduces trauma and saves lives.

The leaders of planet Earth now face the decision of what to do about preparing for contact. It is just as real an issue as nuclear weapons or environmental pollution, but it is not treated that way, which may be a serious mistake.

The Contact Impact Model restates our assumptions about contact with extraterrestrial intelligence. However, it cannot fully describe the qualitative impact of contact on the most fundamental issue of all—how human beings think about the nature of the universe and our place in it. That level of thinking is described in various ways: as "worldviews," "mindsteps," "paradigms," and "cosmologies."

Regardless of the term that is used, it refers to a context in which all other thoughts are generated. In systems terms, there can be enormous changes in the *content* of a society's thoughts, but the basic structure doesn't change until the social paradigm, the context of the thoughts, shifts. When the context itself changes, then there is a fundamental *transformation*, allowing the society to evolve to a new level.

Steven J. Dick, Gerald Hawkins, Michael Michaud and others have argued that the real impact of extraterrestrial contact will be at the cosmological level, and not change but *transformation* will result. They see SETI as another step in the evolution of human consciousness, with key milestones along the way. To talk about how we can get ready for SETI, then, we must see contact from the broadest possible perspective.

Level Three: Impact of Intelligent Life on the Universe

The paradigm shift required to get ready for SETI means moving beyond concerns about the impact of contact on human society, and even beyond our impact on extraterrestrial societies. It may take hundreds, thousands, or even millions of years for terrestrial and extraterrestrial civilizations to absorb the initial shock of contact, but that is a short period of cosmic time.

Planetary unity is not an automatic result of contact, and neither is interplanetary unity. Eventually, however, terrestrial and extraterrestrial societies should realize that they are united by the common experience of being alive and intelligent in an evolving universe, and that working together is the obvious next stage of their own evolution. The key to preparation is to comprehend the potential impact of intelligent life, both terrestrial and extraterrestrial, on the universe itself—that is the context in which unity is realized. While this may seem to be a large conceptual leap, it is nevertheless the most likely long-term outcome of universal evolution.

If our understanding of the evolution of the cosmos is correct, all the matter and energy of the universe were once compressed together into a "primeval atom," which exploded (the "Big Bang"), expanding outward and organizing itself into the galaxies, stars and planets that we observe today.

In our own local region, the Earth existed for millions of years without life as we know it. The vast majority of atoms making up the planet were organized into very simple combinations, primitive information processing systems. One day—we still don't know exactly how—life came into being and began to exercise a tremendous influence on the planet. Living things are a very different way to organize atoms, and a far more complex way to process information. Once organic cells appeared, the Earth began to take a different course from all the other planets in the solar system.

According to one theory, known as the Gaia Hypothesis, life, in the form of the biosphere, manipulates the Earth's systems to optimize its own survival, adjusting to changing conditions over time. For example, even though the sun's output of energy has increased several times since the appearance of life on Earth, the mean temperature of the atmosphere has never strayed far from the levels that are comfortable for the survival of living organisms.[26]

Another day—we still don't know how—intelligent life as we know it appeared on the earth. Intelligence is an even more complex way to organize matter and energy and process information, and intelligence takes up where life leaves off in exercising control over its environment.

For eons, of course, the planet was the whole world for human beings, and its impact on us was far greater than our impact on it. At some point in the recent past, however, a threshold was crossed, and today, the human impact on Earth is so great that we threaten the biosphere and ourselves. We must now learn the discipline of planetary management in order to survive, and we must leave the Earth in order to thrive.

Human beings are already conducting experiments in the creation of new "biospheres" that will allow us to take entire environments with us into outer space.[27] We are also seriously discussing the emerging science of "nanotechnology," which will enable us to manipulate matter at an atomic level, building machines that we cannot even see.[28] The science of "terraforming," the intentional alteration of planets to support human life, is conceptually well advanced, and may be attempted on Mars.[29] Using extraterrestrial materials to build large-scale habitats in free space is a subject of ongoing research as well.[30]

These are the visions of a Type I civilization aspiring to become a Type II civilization, and more. A species that can imagine reconstruction of a planet (terraforming), and of a solar system (the Dyson Sphere), will eventually think about galaxies and the universe itself. It has taken humanity 4.5 billion years to emerge out of the matter/energy flux of the solar system to a point where we contemplate controlling the forces that birthed us. Is there any doubt that our spacefaring descendants, 4.5 billion years in the future, will be any less ambitious than we? Won't their visions be enhanced, and their timetables shortened, if they meet with other species of like mind?

The real question about long-term impact, then, is, "What role will intelligent life, regardless of its planet of origin, play on a cosmic scale?" The next step for those interested in developing the Contact Impact Model is to consider how terrestrial and extraterrestrial intelligence will interact with each other and then with the universe as a whole we must also bear in mind that the space development movement will catalyze the evolution of terrestrial intelligence into extraterrestrial intelligence within the next few centuries.

It is worth noting that the debate about abundance and scarcity of life and intelligence changes when people look into the far future. People on all sides of the SETI question tend to think similar thoughts when they contemplate the fate of intelligent life in the universe. For example, Barrow and Tipler envision intelligence evolving a form of immortality as it gains increasing control over the resources of the universe.

They have defined a "Final Anthropic Principle" in the following way:

> Intelligent information-processing must come into existence in the Universe, and, once it comes into existence, it will never die out.[31]

This vision of Barrow and Tipler still does not necessarily assume many seed-points of life and intelligence in the universe; there may be only one or a few. They consider it unlikely, however, that intelligent life will have been brought into existence by the universe and then allowed to die out before it has had a measurable impact at cosmic levels of reality.

Sagan and Shklovskii also speculate on the long-term impact of intelligence at a cosmic level, saying:

> Cybernetics, molecular biology, and neurophysiology together will some day be able to create artificial intelligent beings which hardly differ from men, except for being significantly more advanced. Such beings would be capable of self-improvement, and probably would be much longer-lived than conventional human beings.[32]

They ponder the future possibilities and ask, "Could a highly developed civilization—one which we have called a Type III civilization—change the characteristics of entire galaxies?"[33] According to them, "Intelligent life in the cosmos tends to have an active influence on the character of the cosmos. . . ." and "We are only entering the cosmic era. What will the future hold?"[34]

Michael Michaud envisions an expanding role for humanity as conscious beings if we embrace the "extraterrestrial paradigm," which provides humanity with a set of transcendent and inspiring goals. By extraterrestrial paradigm, Michaud does not mean SETI alone, but the expansion of humanity off the planet to become a multi-planet species. It is consistent with an "evolutionary paradigm," which assumes that "the Universe will continue to evolve to higher levels of organization, and possibly awareness."[35]

According to Michaud, the most important task facing terrestrials and extraterrestrials is to understand how intelligent life can survive in a universe that is running down as it depletes all its energy sources. As entropy increases, the universe may eventually face what scientists call "heat death," a condition of eternal equilibrium.[36]

Michaud believes that if the evolutionary paradigm can be fully realized by many different cooperating intelligences, the impact would be enormous:

> The end result would be a conscious universe, the ultimate macro-intelligence, supplying the point of its existence and determining the fate of the universe.[37]

Many writers, approaching the SETI issue from radically different perspectives, are led to see impact as far transcending the initial impact of contact on humanity. In fact, the initial impact of contact or non-contact should be seen only as the stimulus that will produce a major response by our species to its larger environment.

In tracing the evolutionary impact of "the Overview Effect," which only tangentially considered SETI, I also concluded that the eventual result of human space exploration would be a "universal overview system," characterized by the universe becoming fully aware of itself:

> As more overview systems are created and linked together, the final outcome might be for the universe itself to become the ultimate overview system.[38]

This line of thought also lays the foundation for a new view of the universe, which I have described elsewhere as the "Cosma Hypothesis."

Can the impact of intelligent life on the universe be described in a formula, similar to the first two levels of the Contact Impact Model?

The answer is a tentative yes, and the formulation for it may be deceptively simple. As intelligent life expands beyond its planet(s) of origin into outer space, populations will presumably grow, and impact will increase in proportion to that growth. In fact, an increasing percentage of the matter in the universe is now being used to create intelligent organisms. The impact of intelligent life can therefore be understood as the ongoing transformation of the physical

universe into a mental universe, a process of "cosmic encephalization" that may be measurable.

The measurement requires comparing an estimate of the total
amount of matter in the universe to the quantity organized into
intelligent life as we know it. The assumption for this level of the
model is:

> The impact of intelligent life on the universe is directly proportional
> to the quantity of matter in the universe organized as intelligent
> organisms.

A catchy phrase captures the idea by stating that the formula for
the impact of intelligent life on the universe is described as "mind
over matter." Specifically, the equation is:

Impact = Atoms organized as intelligent life / Total atoms in universe
Impact = Matter organized as mind (M) / Total matter in universe (m)
$I = M / m$

It is possible to add specific quantities to this formula. There are
some 10^{80} atoms in the observable universe. It can also be calculated
that all of the human beings on earth are organized as 25×10^{34}
atoms.[39] If humans are the only examples of intelligent life, as we
know it, in the universe, then the equation would read:

Impact of Intelligent Life on the Universe = Atoms organized as Mind/
Atoms organized as matter
Impact = 25×10^{34} atoms/ 10^{80} atoms
Impact = 25×10^{-46}

As we might expect, the measurable impact of intelligent life on
the universe, as measured by humans on Earth, is miniscule. The
answer is a decimal point followed by forty-six zeroes and a twenty-
five. While this is a small number, it should be remembered that
several million years ago, it might have been zero.

Once this equation is introduced, it offers the opportunity to
examine a number of questions, and may be the essence of a new
paradigm. From one perspective, it measures the impact of intelli-

gent life on the universe, but it can also be said to measure the level of intelligence of the cosmos.

The Drake Equation takes on new meaning as well. Once it has been used to calculate the number of civilizations in the universe, those figures can be added to the impact equation for a better estimate of total impact. The more civilizations, the greater the impact will be.

"Impact" transforms itself into something different at this level of description, and provides a new way of looking at the debates over SETI. The Assumption of Mediocrity states that the atoms in our galaxy are the same as those of a galaxy millions of light-years away. The model says, "Yes, but how those atoms are organized is the critical factor that defines uniqueness." The Anthropic Cosmological Principle says that the conditions of the universe must be such that observers come into being. The model says, "Yes, and the atoms uniquely organized as observers are still a part of the universe."

Observers are important, but not on their own. As Lynn Harper puts it, "Because of us, the universe is alive; because of us, the universe is intelligent."[40] Observers are really participant-observers.

Intelligent life is not separate from the universe, any more than a brain or an eye is separate from a living organism. The cosmos has created the conditions under which matter eventually organizes into mind, and then questions the nature of the universe and of itself. We do not have to posit a divine creator or underlying purpose to see that intelligence, by its very nature, is driven to create a purpose of its own. By definition, that must become at least a part of the purpose of the universe.

And that definition of purpose is the challenge for all intelligent life, regardless of its point of origin. That should be the topic of conversation among advanced technical civilizations.

Long-Term Impact: Summary

Wide-ranging thinking is directly relevant to the question, "How do we get ready for SETI?" If human beings hunker down on planet Earth, and worry about the impact of contact on an earthbound life, then the experience of contacting extraterrestrials will cause fear and insecurity.

However, if humans think of contact as a welcome step in a much larger evolutionary process, then success will be greeted with joy and expectation. Humans need to consider our ultimate role in a universe larger and more complex than a single planet. We must ponder how we can become "Citizens of the Universe," offering more to the larger whole system than we take from it. Being citizens of the universe means searching for other citizens and finding ways to work with them on common problems of intelligent life. That is the true promise of SETI.

In examining existing assumptions about contact, we discovered that most of them do not hold up to careful examination, and that nothing is automatic when it comes to the results of contact. In particular, none of the supposed benefits are inevitable; we must work to create them ourselves.

"Getting ready for SETI" is a valuable endeavor, regardless of whether current listening programs are successful. In terms of impact, it may also be essential to the survival of humanity and to the taking of our next evolutionary step.

Chapter Ten

Getting Ready for SETI

> We are the local embodiment of a Cosmos grown to self-
> awareness. We have begun to contemplate our origins:
> starstuff pondering the stars; organized assemblages of
> ten billion billion billion atoms considering the evolution
> of atoms; tracing the long journey by which, here at least,
> consciousness arose.

> —Carl Sagan, Cosmos

"STARSTUFF PONDERING THE STARS"—AND NOW, WE ARE REACH-
ing out to those stars as well. The scientific search for
extraterrestrial intelligence holds out the promise of opening up
unimaginable horizons for humanity, answering questions that have
challenged us throughout our existence.

The fascinating truth is that different answers to the question,
"Are we alone?" lead to similar conclusions about what ought to be
done. On the one hand, if earthlings are going to become part of a
universe teeming with intelligent life and a diverse collection of
alien cultures, shouldn't we unite as a planet to maximize our
success?

If the answer is that we are alone, and we will expand outward
into a universe where Earth is a seed-point for life and intelligence,
aren't we also called upon to perceive a higher vision and purpose
for ourselves?

Regardless of the answer to the SETI question, humans must
confront what it means to become "citizens of the universe." In *The
Overview Effect*, the process is described as the "Human Space
Program," a global commitment to exploring the universe as a tool
of human and universal evolution. The Human Space Program re-

sembles Michael Michaud's description of "the Grand Strategy." Common to both is that humans ought to see all forms of space exploration—manned missions, planetary probes, space settlement, and SETI—as part of a conscious decision by humanity to manage its own evolution.[1]

All forms of exploration are journeys of self-realization for individuals and for society. A search is a quest, both for what is "out there" and what is inside. Because this connection exists, the evolution of the universe, the species, society and invididuals are really one process.

SETI is explicitly a search, a pure form of exploration. It is part of the ongoing reconnaissance that our species is conducting as we take our next step off the planet and into another evolutionary epoch.

An assertive Human Space Program, pushing the boundaries of human thought and presence outward into the cosmos, is a message that says, "We want to evolve. We want to see new sights, think new thoughts, live in new environments." However, it need not be an aggressive, imperialistic process. The SETI component of the Human Space Program says, "We know there may be others out there. We seek friendship and cooperation, not domination."

There are those who criticize the space settlement component of the Human Space Program. They say that human beings have not learned to live within limits, and feel that humans ought to accept limits to growth, learning to work with their home planet, before venturing into the solar system and beyond. However, human evolution does not have to be antithetical to ecological concerns. Conscious cooperation with the universe is the most important issue that intelligent life faces, and it is nothing less than a cosmic ecology project. Without living systems, it appears that the universe will run down and collapse. With living systems, some sectors of the universe will survive, perhaps long enough to learn the secret of cosmic immortality.

The universe is still evolving; it is not a finished thing, and we don't know much about its true nature. Can it be that life and intelligence are an evolutionary strategy developed by the universe to ensure its own survival? Intelligence allows for learning, and there is no inherent reason why human beings cannot learn to live in harmony with their environment, on one planet or many.

These are difficult questions, but they also raise the stakes of SETI. If we take our minds off the more limited issue of how contact will affect our current planetary society, we might ask, "Do the extraterrestrials have any clues about the destiny of intelligent life? How much more do they know than we?"

Getting ready for SETI, then, involves thinking about important questions regarding human purpose in the universe, a familiar topic, now played out on a much larger scale than in the past.

How to Begin

Now that the framework of getting ready for SETI has been developed, let's look at specific practical steps that ought to be taken.

Preparing for contact should be seen as a global process and a global priority. As part of the International Space Year in 1992, a global conference on SETI should be held to discuss a planetary strategy for SETI, including scientific, political, economic, sociological, philosophical and psychological issues. The conference should consider new approaches to detection, especially whether phenomena not yet explained by natural causes may be the product of intelligence. In addition to the protocol question, this conference ought to consider how to respond to a signal, as a first step in developing a comprehensive policy foundation for interstellar relations.

Broadbased education about SETI ought to be expanded. The outreach efforts of the NASA project, the *Contact* organization, Planetary Society, and of other organizations deserve support. A major foundation or corporation ought to make SETI education one of its primary objectives.

A major international SETI media conference should be held as soon as possible. This conference can either be held independently or as part of the 1992 conference mentioned above. Its purpose would be to educate the media about SETI and prepare them for responsibly dealing with reports of successful contact.

An inventory of current SETI activities should be created, including a computer network of experts to support the media in their reporting on SETI and possible contact: The network can also become a resource for the SETI professionals, helping them to learn about new ideas and approaches and to avoid duplication of effort. The inventory will also be extremely helpful to journalists in developing current and future stories of high quality.

The philosophical dimensions of SETI ought to be explored as well as the technical aspects. The philosophical and humanist aspects of contact are what interest most people, and it is in this domain that the most important results will be felt. A major university ought to start an institute for this purpose, or perhaps the SETI Institute, *Contact*, or others should begin research in this area.

An interdisciplinary team of physical and social scientists should begin an intensive investigation of culture contact and species contact on Earth. We need to understand what made those contacts successful and what made them difficult. We can learn a great deal from both sides of culture/species contacts, and the insights of this study can then be applied to SETI.

A contact verification committee should be established. The committee can play a positive educational role now, and will be ready to perform an even more important function if contact occurs.

We can start all of these projects, and more, within a few years. Everything is in place for a Human Space Program to begin, if it is seen as comprising the sum total of all activities now taking place on Earth, and is oriented toward human evolution into the universe. Everything is also in place for a global SETI project to begin as a subset of the Human Space Program.

Human beings consistently confront issues subconsciously even if they are not aware of them on a conscious level. As regards extraterrestrials, the preparation for contact is well underway within the popular culture. Many of the possible contact scenarios are being played out in books, on television, and in films.

These explorations are positive, because they allow people to think about extraterrestrials in a variety of ways. What is needed now is a more sophisticated approach that considers the ambiguous impact contact might have on human beings. Carl Sagan's fictional account of a successful SETI project (Contact) goes far in that direction, and much more is needed.

In another domain, those who dismiss UFOs as "unreal" might find the phenomenon more understandable if they considered it as another way of getting ready for contact. As with films and television programs, the UFO contacts range from terrifying abductions to calm interactions with kind extraterrestrials who are concerned about our ecological problems, and who are humanoid themselves.

Is there a lesson for us in the experience as we move into the new phase of a scientific search for extraterrestrial intelligence?

The people of Earth are planning for contact, some more consciously than others. The trend of interest in extraterrestrial intelligence will continue because it taps into something deep and profound in human nature. Does this mean that human beings are "ready for SETI?" The answer is, "probably not." Significant government funding of the search is a recent phenomenon, and the involvement of social scientists, philosophers, and other non-technical people is relatively new. The planet as a whole is focused on more immediate problems, and contact preparation is a low priority.

However, there is time available, and the stakes are high. What would have happened to native American society if they could have prepared for the Europeans coming to the shores of North America? It might have been infinitely better for both cultures if forethought had informed the process.

Human beings can exercise forethought now. We have the capability to think ahead and plan, to control our destiny. We also have free choice, and having chosen to go forward with SETI, we should now choose to make it work for ourselves and the "others."

At one level, the search is aimed at finding the answer to the question, "Are we alone in the universe?" and at another level, the question is, "If there are extraterrestrials, what are they like?" However, the question is really much deeper and more personal than that. It seems to really be "Who are we?" or "Who am I?" As a species, we are claiming to be mature enough to compare ourselves with others on a cosmic scale. By trying to find extraterrestrials, we are really searching for a deeper understanding of terrestrials.

In the words of Lynn Harper:

> What does it mean to be human? Although philosophers have grappled with this question for eons . . . we have never had a context within which this question could be meaningfully addressed. . . . Now we can pose the following questions *because we may have to grapple with the answers:* "What does it mean to be human if humanity is the only intelligence in the galaxy or in the universe?" and "What does it mean to be human if there are other technologically competent beings in the universe?"[2]

What *does* it mean to be human? Each of us has our own answer; you don't have to be an expert to respond to that question. It means being happy and being sad, loving and hating, writing poetry and grocery lists, looking up at the stars while trudging through the mud. It's hard being a human, constructed equally of the mundane and the sublime. We have always been sharply aware of the struggle of matter becoming conscious, of life in two different realms. But in that we are not alone; it's a universal struggle.

Maybe it's hard being an extraterrestrial as well. Perhaps when we meet, our species will have great tales to tell as we sit around the cosmic campfire, about how we emerged out of ignorance and darkness into the light of intelligence and awareness—and of the price we have paid for our sentience.

The immediacy of the situation makes it compelling and important rather than just speculative philosophy. Now we must bring these questions up out of the collective unconscious and directly into the light of day. We may need to find common ground with the others sooner than we ever imagined.

Getting ready requires a massive involvement including everyone on the planet, not just a few "experts." It has implications for every "Terran," and not just the humans on Earth. Humans are one part of a process that has taken place on this planet, but our fate is intimately bound up with that of all other life forms here.

If SETI implies asking the question, "What does it mean to be human?" it also means thinking about what it means to be a life form originating on Earth. It is not just humans who are evolving outward into the larger universe, but the biosphere of Earth itself. If the process is being replicated elsewhere, it means that the biospheres are radiating out toward one another, creating evolutionary possibilities beyond imagining.

The view of Earth from orbit and from the moon produced the "Overview Effect," a realization of the unity and interconnectedness of our planet as a whole system. The SETI Factor is bringing us to an advanced version of the Overview Effect—a realization of the unity and one-ness of everything in the universe.[3] When and if we make contact, we ought to consider how rare and precious life and intelligence are in this vast and ancient cosmos. Within that context, there are no "aliens," only "Us."

Appendix A

Interview with Dr. Isaac Asimov, Author
August 11, 1989

WHITE: What is your current thinking about life and intelligence in the universe?

ASIMOV: I think that every star has planetary bodies. There is no way of telling what percentage are sufficiently Earth-like to be hospitable to life as we know it, but considering how many stars are in our own galaxy, it stands to reason that there must be a large number of Earth-like planets. I believe that on every planet that is Earth-like, life will develop.

The problem is, what are the chances of *intelligent* life developing? There is no way of telling; we have absolutely no evidence. We have ourselves, but we don't know if the course of evolution on Earth has been typical or not. Even if it has been typical, the fact remains that life existed for three and a half billion years on Earth before a technological civilization arose.

These civilizations may be few and far between, and it is not inconceivable that we might be the only one. Furthermore, not long after we developed a technological civilization, we have reached the point where we are perfectly capable of destroying our civilization and ourselves as well. Every time a technological civilization develops, it may have to withstand the chance that it will destroy itself, because it will quickly gain enough power over the environment to be able to use the laws of nature to self-destruct.

Perhaps this happens routinely, and the galaxy and the entire universe may be absolutely littered with planets containing life that is on its way to developing technological civilization, and planets in which the technological civilization has come and gone in the blink of a geological eye. Perhaps only on the rarest occasion will we find a technological civilization in that brief period of time between its formation and its destruction.

WHITE: How do you feel about the current strategy of searching for extraterrestrials in terms of searching for signals? Does that make sense?

ASIMOV: It's the only thing that makes sense. There is very likely no life of any kind in the solar system except on Earth, so if we're going to find examples, it will have to be at stellar distances. We have to ask ourselves, "What can cross stellar distances?" Short of actually transporting intelligence, how else could we detect civilization elsewhere, and the obvious answer is radiation of some sort.

The only forms of radiation that can reach us from other planets without being hopelessly garbled in the process are such things as photons, gravitons and neutrinos. The latter are difficult to detect, so it makes sense to work with photons. Visible light won't penetrate dust clouds, so we have to try to detect something in the radio wave region. The shorter the better, since they are easier to detect, which gives us microwaves.

Any technological civilization will learn how to receive those particular wavelengths if they want to learn anything at all about the universe. If they can receive it, they must also know how to emit it, so that again seems to be the logical approach. We're attempting to study the universe in detail, looking for microwaves that are not random in nature nor regular. Neither randomness nor regularity carry information. We need something that is irregular, and not random. That would indicate the presence of intelligence, especially if we could find in the irregularity certain meaningful regularities.

Even if we do detect signals that seem to be of intelligent origin, they may not be aimed at us. They may just be natural spill-overs from that civilization's communications. What good does it do, then, if the signals are meaningless to us?

The answer is that they do send the following message: "We are here." We know then that there is an intelligent form of life, we may be able to pin it down to a specific stellar system, and we know that technologically they are more advanced than we are. From that, we can conclude that it is possible for civilization to attain a technology superior to ours without destroying itself. I can't think of a message more important than that, because we need hope.

WHITE: I've been fascinated with the historical interest of humanity in extraterrestrial presences.

ASIMOV: Yes. It's only very recently that human beings decided they were *not* accompanied by other forms of life. Most people were not only certain that human beings were not the only intelligent creatures in existence, but that all the others were far more powerful than we are. They believed in the literal existence of angels, fairies, spirits, everything.

Then they took for granted that if the moon were another world, it must have people on it, perhaps not very different from us. That is the attitude we see on television. We find many aliens on "Star Trek" and similar programs, and while they are different enough from humans clearly to be another species, they are still recognizable as primates.

WHITE: Have you seen changes in the way science fiction deals with aliens?

ASIMOV: Yes. A great deal of credit for that goes to "Star Trek." They have malevolent aliens such as the original Klingons, but they also introduced the Vulcans. Then *Star Wars* had those cute little robots, and *E.T.* presented a child-like extraterrestrial, which is completely sympathetic. The movie people found out that good aliens are good box office.

WHITE: I see it as a maturing attitude, being less afraid and no longer projecting our own fears onto the universe.

ASIMOV: I hope you're right. Our experience rests in the European exploration of the rest of the world, in which we enslaved the natives we found, and then killed them off. We expect the aliens to be as bad as the Europeans were, but even we have now learned that it isn't right to kill off natives or even an endangered species.

If we can learn, I imagine others can as well. I think that any technological civilization composed of highly combative people can never combine together long enough to get out into space. Exploring space is such a tremendous undertaking that it can't be done by nations that spend all their time planning war against each other.

Therefore, I think any civilization that can explore over interstellar distances must have overcome their petty combativeness, and they will be, by human standards, benevolent creatures.

WHITE: There may be a galactic civilization, and we may be getting ready to enter into it.

ASIMOV: Yes, and we can argue impact in two ways. On the one hand, we can learn from them. In the space of a generation, we will gain enough knowledge and insight to bring us to a point that we might not reach by ourselves for a long time. On the other hand, we would be deprived of the fun of doing it for ourselves, and we might even suffer a bad case of inferiority, realizing we are just not bright enough to keep up with these others.

I think we would learn. And no matter how much we learn, we can still have the fun of finding out more. Also, the extraterrestrials might conceivably protect us, establishing a kind of reservation to allow us to continue developing on our own.

WHITE: What about the impact of not making contact?

ASIMOV: There could be a new kind of loneliness and desolation, developing a fear of a vast impersonal universe in which we are lost. The other side of the coin is that we could begin to appreciate the uniqueness and preciousness of Earth, and save it not only as our home, but as the only home of an intelligent species in the entire universe.

WHITE: We might begin to have a higher sense of purpose because Earth would be the seed planet for life and intelligence.

ASIMOV: Yes, and now it's up to us to colonize the universe and make it a common world for life. Another reaction might be to prove that Earth is a special creation. It might create an enormous religious revival.

WHITE: In contrast to the Assumption of Mediocrity, which says that nothing extraordinary happened here, and it's probably happening everywhere. What about the other school of thought, the Anthropic Cosmological Principle?

ASIMOV: The idea is that you can't argue that there must be many forms of life because of the size of the universe. You might need an extremely large universe to supply the kind of properties making it possible for life to develop on one world. But having achieved that, it doesn't mean that life will develop only on one world. In order for a raindrop to form, you have to have some concatenation of clouds and wind patterns and temperature change that will involve a large section of Earth. But one raindrop isn't the only rain that falls; billions of raindrops fall, and I think that if the universe were large enough for one world to develop life, it might well mean that many worlds could then develop.

Interview with Dr. Ben Finney, Anthropologist, University of Hawaii August 20, 1989

WHITE: Could you discuss the motivation behind SETI as compared to people interested in interstellar migration and space development?

FINNEY: There is an apparent contrast, and I've puzzled over that. I'm more interested in the human presence in space than in extraterrestrials, but I intellectually recognize that you have to think about the possibility of extraterrestrials. When I was working with John Billingham at NASA, he put me in the SETI unit, so I was always trying to reconcile the two.

There were people in both camps who were dead set against the other, but I felt that a lot of the difference may simply be one of taste or experience. I always had a feeling when I was working with the NASA SETI group at NASA/Ames that I was with a group of "superhams," who wanted to sit in their armchair and fiddle with radio circuits to get that ultimate connection. They said that information is the primary good and physical experience is secondary, and therefore they were after the higher good. The space colony people seemed to be much more visceral and experientially oriented. They wanted to see, taste, and feel things.

In a paper that I presented in one of the international astronautical federation congresses, I tried to point out that, "Well, maybe the two camps are not that different after all." They are both expressions of an exploratory urge, one more sedentary and the other more active, but nonetheless expressions of something very similar and in the end, they have similar scenarios—a galaxy populated by intelligent civilizations, though by different sources.

WHITE: If the universe is already heavily populated, that is one clue to the space development people's lack of interest. The motivation of many space development people is in unfettered human expansion, a liberation from the restrictions present on a highly civilized Earth. For those involved in that movement, there may be a hope there aren't too many extraterrestrials, or our scenario has to be modified.

FINNEY: Yes. A number of people, if pressed, will have to logically admit the possibility of extraterrestrials, but they would rather say, "Well, extraterrestrial civilizations must be pretty sparse and we will probably have to go a long way into the galaxy before the issue is faced," and then they forget about it. The SETI people are doing just the opposite. They hypothesize a galaxy heavily populated with extraterrestrial civilizations, and are betting that we may be able to detect their radio transmissions.

I don't put much store in our a priori ability now to say yes or no. It's simply an experimental question, and we have to look. But I don't think the presence or absence of extraterrestrials is going to be a pressing issue until we expand out into the solar system. Once we've done that, then we are going to be looking at other solar systems, and then it will be a most vital issue, because it will be a biological concern to our species—"Do we have room to expand further or don't we?"

Given our present state of cultural evolution, efforts to solve that question will attract much more support (including financial) than currently is flowing into the resolution of the grand intellectual puzzle of "Are we alone?"

WHITE: When it becomes an actual expansion issue that affects people's lives on a daily basis, then we will really want to know?

FINNEY: Then there will be a lot of resources poured into it. Now, the SETI people feel they are doing science for science's sake with some overtones of helping humanity. Similarly, if I talk with some of my colleagues in planetary science, who are sending spaceships to Mars, Jupiter, or wherever, they will say, "This is basic science. What you are interested in is science fiction." I would say that both the planetary scientists and the SETI people are really performing a reconnaissance for the species for future expansion. In a sense, they are the unwitting tools of the species, intelligent life, or DNA itself. They believe they are doing science, but they may be doing something else.

WHITE: You're looking at it as an anthropologist.

FINNEY: Yes, I look at it as, "We're an expansionary species, and there is a division of labor." We are an inquisitive, exploring animal that has spread over the globe and is now beginning to expand into the solar system.

WHITE: Regarding impact, Michael Michaud mentioned to me you had the idea that the farther away the other species is, the less disturbing it will be for our species.

FINNEY: I was reacting to pronouncements of impending doom if we make radio contact with an extraterrestrial civilization, and (in contrast) the cargo-cult-like prophecies of all the wonderful things that will come to humanity through learning about and from an exotic form of extraterrestrial or intelligent life. I was expressing a degree of skepticism as to whether we would be able to make any sense whatever about what a radio transmission from elsewhere in the galaxy might mean.

After the initial excitement, there will be a long haul of trying to figure out what anything means. I can't see any great or immediate impact other than the realization that we are not alone. And we may have already "discounted" that, as they say in the stock market. It may be that the younger generation, having seen E.T. and all of these movies, may say "Well, sure," and it won't be a great philosophical reversal. We have plenty of examples of culture contact in which people completely misunderstand each other, and that's within the same biological system, indeed the same species. My example is the contact between the Europeans and the Melanesians that generated the cargo-cults.

The Europeans showed up with all these marvelous "goodies of industrial society," and the New Guineans attempted to understand this through their conceptual system that says it's their ancestors who generate material technology and give it to the people. Here you have completely new forms of technology, so therefore their ancestors must have made them, and somehow the white folks have intercepted these technologies, and kept them all for themselves. So the New Guineans think that if only they do the proper rituals, then they can get direct access to all these goods—"cargo" in their terminology.

Well, they don't understand our system of manufacturing and all that, and we've spent fifty years trying to figure out their cargo cults, and what is going on in their minds. Yet, here we are members of the same species.

WHITE: And we don't understand them. . . .

FINNEY: We don't, and here is where I part company with some of the SETI people. Let's say we get a transmission featuring prime numbers, Planck's Constant or some other mathematical verity. From that, some SETI people assume we can begin to learn the history and culture of the extraterrestrials, which will bring us all sorts of revelations, such as how to stop nuclear war. I don't see how they make that leap. Phil Morrison has written more sensibly than anyone in the SETI camp on this. He says there will be whole new industries formed around the long, slow and difficult task of the interpretation of the messages.

WHITE: I recently went to a conference on "Native Science," and I realized that we have been interacting with that belief system for three hundred years, and we don't understand it.

FINNEY: Yes. The history of science tends to discard everything that doesn't lead directly to Western science. In anthropology, we are deliberately dealing with everything that *didn't* lead to modern industrial society.

WHITE: When people talk about the impact of contact, they assume there will be a very dramatic impact, perhaps very devastating to us if we meet a "superior" culture.

FINNEY: I can imagine headlines and so on, but what do you do after that if you don't know what it means? It may become a non-story very quick, and if interrogations take light-years, it could settle down into an intellectual industry of trying to figure out what it all means. Maybe the impact will be an initial flash, then a slowdown, and then a slow percolation. But I don't see disaster from discovering that we are inferior in the universe, because it won't be readily apparent that that is the case.

WHITE: The industry's task might be to figure out where we are in comparison to them, to see where we fit in evolutionary terms.

FINNEY: Yes. My colleagues in SETI seem to take the point of view that science is inevitable, and that radio astronomy is inevitable. I wonder what kind of civilizations we might have that didn't go that way.

WHITE: One of the motivations in doing SETI is that if we get a signal, it shows that technological civilization will survive. Based on our own history, however, there is also a feeling that an advanced culture will destroy a less advanced culture. Do you have any ideas about that?

FINNEY: Well, you have Europeans expanding to Asia, the New World, Africa, and the Pacific. But would you say they destroyed China?

WHITE: No, they had a big impact on it, but didn't destroy it.

FINNEY: Did they destroy the Hawaiians?

WHITE: No.

FINNEY: As a viable culture in the ancient sense, yes. It's greatly transformed. There are differences in scale, of course. The cultural and technological impacts have been very great and imported epidemic diseases almost wiped out the Hawaiians. But now there is a growing part-Hawaiian community made up of people who adhere, in part at least, to Polynesian values and ideals.

Interview with Dr. Paul Horowitz, Harvard Professor, Developer of Project META October 31, 1989

HOROWITZ: It's important to understand that any civilization we contact is guaranteed to be more advanced than we are; civilizations even slightly less advanced can't play this game at all. They're also guaranteed not to look like us, i.e., upright bipeds with stereoscopic vision.

WHITE: If we're merely listening, then it seems almost impossible that we're going to hear from some species that's not more advanced. As for how they'll look, you seem to be saying that they'll be radically different.

HOROWITZ: There are some people in this business who'll disagree with me. Dale Russell, a Canadian paleontologist, did an interesting study of the most advanced dinosaur of its time. He simply turned the evolutionary crank for another sixty-five million years and produced something which is remarkably like us, although it's cold-blooded, still a reptile.

I don't think there's much agreement, however, in the paleontological community about these issues. I think the main point of view is that quirky things happen in evolution. This quirkiness leads some people to throw up their hands and say, "We could never have another advanced civilization like our own."

I think the more reasonable point of view—the kind that Steven Gould (professor at Harvard) would give you—is that being smarter is a good adaptation and you're likely to have mutations that lead you to handle the environment better. It's the ability to process information within the environment and take care of yourself that matters. It doesn't matter what "shell" it's in. The extreme point of view there would be Fred Hoyle's book about the "Black Cloud." From a single planet, we have a pretty wide range of life on Earth. You multiply that by powers of ten, and are you going to get warm-blooded, two-legged people? Given these factors, the extraterrestrials may look unappealing, and people on Earth do pay attention to looks. There's probably a real shock in store there.

WHITE: It's interesting that E.T. is now a dominant image of extraterrestrials, and though he appears to be different at first, he is not really very different from us.

HOROWITZ: He could happily have evolved on Earth . . .

WHITE: In fact, he breathes our air . . .

HOROWITZ: He's got a schnozz (nose), and he's got the same number of legs and hands, and he has fingers . . .

WHITE: Does all of this suggest that radiotelescope searches might not work for such creatures?

HOROWITZ: Radiotelescopes, as efficient communications devices, are really a consequence of conditions in the galaxy, not of the particular structure of the human animal. Alien radiotelescopes probably look a lot like ours. Now, perhaps we would have difficulty building radio telescopes if we were dolphins, but I don't think dolphins are going to be able to communicate any other way, either.

I guess we're narrowing our search to civilizations that are not only smart, but also are able to construct the technological paraphernalia to communicate. But they don't have to have two hands that match, and if they go around on wheels, that's okay.

WHITE: You once wrote that, based on the searches done so far, we know the galaxy is not teeming with life and intelligence that is trying to communicate with us in straightforward ways.

HOROWITZ: You can pick holes in that argument, but it's a conclusion you can draw.

WHITE: I think there's an assumption on the part of some people that we've been listening for thirty years and we haven't heard anything, so we haven't learned anything.

HOROWITZ: We'd learn a lot more if we found something, because we can't rule out much of anything by not having found it with our very weak searches. With a very thorough search, you can start to rule out plenty of things, but it's been a weak search, so you can only rule out the strongest possible transmission scenarios.

The really intense and persistent irradiation of our solar system by radio signals in the wavelengths that we consider obvious doesn't seem to be happening. It doesn't seem that someone is trying to get through to us in an intense way.

WHITE: What's the current status of Project META?

HOROWITZ: META has been running continuously for four years, searching the northern sky for unusual signals near "magic" frequencies, so far without success. It is the best there is right now, but it's still not very much. Given its constraints, META can do certain things, but it's anemic compared with what you'd like to do. What NASA is working on is much more what you'd like to do, and we'd like to do even more than that. You really don't want to point to what we've done with too much pride. It's really just the first steps in this business.

WHITE: But looking back, don't you think *Apollo 11* will also seem pretty primitive in perspective?

HOROWITZ: *Apollo 11* succeeded. Phil Morrison says the *Nina*, *Pinta* and *Santa Maria* weren't jet planes, but they did the job. META hasn't yet done the job.

WHITE: I know you're planning to do Southern Hemisphere work. Will you continue with the Northern Hemisphere as well?

HOROWITZ: Yes, but we're really planning to go much farther. I think META is under-powered, so we're thinking about expanding the search another hundredfold. We're talking about 100 million channel receivers covering hundreds of megahertz.

WHITE: What do you think about the anomalies picked up by META and other projects?

HOROWITZ: I don't have a good explanation. They're too large to be noise, so they're something artificial. I think there are just curious ways in which interference can sneak through the algorithms meant to reject non-celestial signals.

WHITE: So you think they're Earth-based?

HOROWITZ: Yes.

WHITE: You've been getting about three a year?

HOROWITZ: There were more than that last year. In about two years of running, the list had about twenty-five objects on it. I don't take these things too seriously. A genuine signal has got to repeat and they just never do. They don't even repeat locally. So we're left with a non-repeating thing and not one of them has ever come back. I've checked them all.

WHITE: This naturally relates to another question, which is announcement procedures.

HOROWITZ: I think I could easily be convinced by a spokesman for either side to go along with their point of view because I don't think the consequences are too serious either way. My knee-jerk reaction is that in other fields of science we don't need rules to tell us what to do or how to behave when we've discovered something, and I don't know why this should be a special case.

We have cold fusion situations because some people don't know how to follow up on their own discoveries, but I think that's preferable to having commissions of inquiry going around checking everyone's discoveries.

A false alarm isn't going to hurt in the long run because one way or another, you'll know the truth. The wrong information is not going to be accepted as truth for long.

Interview with Dr. Philip Morrison, MIT Professor, SETI Pioneer September 19, 1989

WHITE: Looking back at the time thirty years ago, when you wrote your paper with Dr. Cocconi, how would you evaluate what we've learned in this first period of SETI?

MORRISON: I'm not so sure either of us had very clear ideas of what would happen when we wrote the paper. We both felt that it would take some time for people to come around to this idea, and we wanted to see what people would say about it. When we put it out, there was a lot of opposition, but there was also a lot of support, about what we thought would happen.

Gradually, the idea has seeped in, until now more and more people are in support of SETI. A rather determined opposition has also sprung up, which I don't mind. As long as they give their reasons, I don't care at all. We are not saying it's highly probable or improbable that we will succeed. The reason for an experimental program is to find out. We said the issue should be turned from the domain of speculation to that of scientific experimentation, or exploration, which might be a better way of putting it. SETI is not based on a single scientific hypothesis, but on many. It is a very complex proposal.

A lot of good radio astronomers have now thought about it, and they've found out various things you ought to do, and they have tried them. We didn't quite anticipate that the power of digital computing would grow so enormously. You can do the whole thing much more easily than we ever thought it could be done, using multi-channel receivers. The searches are being done with ten million channels, which is at least one thousand times greater than we imagined.

That is a positive development. The negative side, which should also be mentioned, is that I was hopeful we would find life on Mars, which would "turn life from a miracle to a statistic." So far, the space program has not produced any evidence that life is not a miracle, and that is disappointing.

WHITE: But people are seeing a lot of organics present in the solar system.

MORRISON: We would have anticipated that in the beginning. That's very positive, but we would have guessed that. Molecules are not life.

WHITE: So when you first made your proposal to search for extraterrestrial intelligence in 1959, you had a sense that organics might be found, there might be life on Mars, and we might be able to pick up signals as well. You didn't anticipate the negative results of the probes to Mars.

MORRISON: That's right. In those days, there was even the claim that there was seasonal darkening on Mars, as though you were seeing growing plants.

WHITE: When I started on this book, I didn't know about the advocates of the Anthropic Cosmological Principle, the people who think that life and intelligence may be incredibly rare. Then there is a middle of the road school that feels life may be common and intelligence is rare.

MORRISON: I try to steer away from those speculations. They are interesting, and they are mirrors of what people bring to it, but we have no knowledge about this, just none.

WHITE: Paul Horowitz wrote an article in which he said that results to date indicate the universe is not teeming with intelligent life trying to get in touch with us.

MORRISON: That's correct, but I think that was already established by Drake's Project Ozma. I think Paul has also established another thing which is that the problem of local interference is circumvented by the technique he uses. He can search so many bands, and they are so narrow and shifting that they are not polluted by interference.

WHITE: I've also been giving some thought to strategies for detection other than radio signals.

MORRISON: I don't think any others are in sight. It's not an impossibility, but none I've heard about are very attractive.

WHITE: My thought was that we might apply Lovelock's ideas on a broader scale, looking for regions of lowered entropy surrounded by regions of increased entropy. It's something like the idea of a Dyson Sphere.

MORRISON: In its day, that was a very valuable idea, but you never do engineering to squeeze out the very last ounce of energy. There are always diminishing returns. If you look at any engineering design, it is not 99.9999999% efficient. No one would go to that trouble—capital costs, sunk free energy—so what you expect is not an object with no light left from the star and a huge infrared signal, but an object with plenty of light signal and plenty of infrared signal as well. But it's much harder to find, so it's not a good method.

WHITE: If a civilization is operating on a scale beyond a single planet, even if they don't build a Dyson Sphere, wouldn't there be an effect we could detect?

MORRISON: Not according to the way people talk about occupying planets today. I don't think the terraforming of Mars and Venus will be done, if it leads to another ten billion population planet. So you won't have ten billion people's worth of energy. You will have a small amount of energy compared to what you've already seen. You see one bright planet and a lot of dim ones. Certainly we need to find planetary systems. That will happen, but that's not ETI.

WHITE: So your feeling is that we're not likely to be seeing any other forms of detection in the near term.

MORRISON: This field is full of speculation, and that's all to the good because it should be. But if you look at the experience, it is not in favor of these speculations, because of the distances involved. We always said, just listen for a time, a long time, maybe a century or two centuries, and then you can start other things. Sending probes doesn't do any good, looking for lots of energy doesn't do any good, and looking for Dyson Spheres is the wrong tack, because I don't think they will exist.

You might look for "Semi-Dyson Spheres." In that case, the people have done their best to capture a star's energy, and have captured half of it, or a third of it, or a tenth of it. To do more doesn't make sense. If Gerard O'Neill is right, then you can make some very convenient habitats out there, but I don't think he is right. I think those habitats will be wiped out by solar flares, unless they're made very heavy and expensive. So there may be colonies, but not with ten billion people. If you want to improve things for ten billion people, you improve Earth. You could do a lot for Earth, more than you can do with habitats.

WHITE: If that line of reasoning is correct, the only way to detect the presence of such a civilization would be communications.

MORRISON: At least the easiest way. Every evidence is that communication grows tremendously, and there's no historical argument that such a civilization wouldn't be very communicative.

WHITE: I'd like to talk a bit about the impact of acquiring a signal. I'm not sure that the media are prepared for dealing with it effectively.

MORRISON: I think that's true, and I don't think it really makes any difference. I've never been of the opinion that the important thing is the first three months, or the first year. The important thing is the slow progress of this understanding throughout the entire intellectual community. It's more like the idea of Copernicus than anything else. What counts about that idea is that everybody now grows up believing that the earth is just another planet.

WHITE: In that case, the impact of SETI would be on the order of hundreds of years, rather than weeks.

MORRISON: That's right.

WHITE: I appreciate the opportunity to talk with you about this. You probably have more perspective about SETI than anyone, from the time you wrote your paper.

MORRISON: I've certainly been thinking about it for a long time, but it doesn't help a great deal to do that. Experience is what counts now.

Interview with Dr. Carl Sagan, Planetary Society Founder, SETI Pioneer February 20, 1990

WHITE: Since writing *Intelligent Life in the Universe*, have you changed your thinking about how many advanced civilizations might be in existence?

SAGAN: Not really. In a way, it's quixotic to imagine that we, who know about only one barely-technological civilization, could tell how many there are in the whole Milky Way galaxy, much less the cosmos. There is a bit of *hubris* implicit in the Drake Equation. But having said that, I would stick by the estimates today that I gave then. If you run through the terms in the Drake Equation, and look at the fraction of stars that bear planets, and the fraction of those planets that are ecologically suitable, we have learned a lot since the mid-1960s. And everything that we've learned is all in the direction of "Planets are a dime a dozen."

That's a very important transition from speculation to data. Something like half the solar-type stars in the neighborhood of the Sun have accretion disks, solar nebulae of just the sort that we believe the planets of our solar system were formed from, five billion years ago.

In addition, the time available for the origin of life on Earth has shortened. The antiquity of life has been pushed back in time somewhat, and the evidence that the Earth, shortly after forming, was inhospitable for the origin of life has been greatly strengthened. So the period of time in which the origin of life can occur has significantly narrowed. Unless you believe that the origin of life on Earth was exogenous, rather than home-grown, then that must mean that the origin of life is now a more likely process than we thought twenty-five years ago.

Those are two areas where I claim the progress of science has made our original (optimistic) estimates much more likely to be right. You might also say that the present very modest trends to reverse the nuclear arms race are an indication that the "L" variable in the Drake Equation (the lifetime of technical civilizations) might be larger. On the other hand, the other catastrophes that have come to the fore since then—global warming, ozone depletion, nuclear winter—work in the other direction. As Shklovskii and I emphasized then, the uncertainty regarding L is the greatest uncertainty in the whole problem, and that remark still applies.

I think I would still stick with some number like a million civilizations more advanced than our own in the Milky Way galaxy.

WHITE: What expectations do you have, with Project META expanding now, and the NASA project going on-line with new computing capabilities? Are we now moving into a second phase of SETI, where it's becoming more entrenched, more acceptable, more heavily funded, and more sophisticated technologically?

SAGAN: The only word to which I would object is "entrenched." It is as superficially entrenched in the world scientific enterprise as I can imagine. But the rest of what you say is right. The technology is still far ahead of our willingness to use it. One clear demonstration of this fact is that by far the most sophisticated radio search program for extraterrestrial intelligence has been carried out by a membership organization, the Planetary Society. It has been paid for from membership contributions, and one sizable gift from Stephen Spielberg.

If META is so much better than all the search programs tried before, and yet is supported by five and ten dollar contributions of Planetary Society members, then it is a clear indication of how much more the government could do on SETI if it wanted to.

It looks as if the government does want to do more, with the emerging NASA SETI program. That will be as big a step ahead of META, as META was above its predecessors. With NASA doing it, it has an institutionalized character, which is very good. It does not depend as much upon the vagaries of continuing support. The Planetary Society does not have a huge amount of discretionary funds, and we will be most happy to conclude our transitional and catalytic role in SETI when the NASA program comes on-line.

As far as the success of META II goes, if you believe that there is only one powerful signal (or a small number of them) that could be detected on Earth, then there is some chance that it can be seen from the Southern Hamisphere and not the Northern Hemisphere—because the source is in the Southern Celestial Hemisphere, and not in the Northern Celestial Hemisphere. So we could be as sophisticated as we wanted to, searching the northern skies and we would miss this bright and obvious signal in the southern skies. That is what META II is about.

On the other hand, if you believe that there must be many such signals, then the fact that we haven't found any in the north is of some significance. However, judging success and failure in this business always involves judging the intentions and capabilities of extraterrestrial civilizations, about whom we have no information whatever.

WHITE: Then there is the question of impact on our society, and theirs. It seems to me that the only variable we can control is our own preparation to participate in a much larger community of intelligent beings.

SAGAN: As I discussed in CONTACT, some people will be made extremely nervous and anxious, because those beings will almost certainly be more advanced technologically than we. We will be revealed as the dumbest civilization around since we have just achieved the capability for radio contact. We won't be able to hear from anyone who is even a little behind us (because they won't have big radiotelescopes), and anybody else is very unlikely to be exactly at our stage of development. If we receive a signal or a message from them, they will already be in our distant technological future.

The idea of finding out what much smarter, much more long-lived, much more capable beings think is worrisome. What if they have beliefs different from ours? It would challenge us, across the board, and not just on social, economic, and religious issues. But in other areas as well—what if we've made some simple, dumb mistakes in mathematics, or physics, chemistry or astronomy?

There are some religious groups that hold tightly to the view that we are the pinnacle of creation, that there are no other intelligent mortals, and there is nothing in the Bible about other civilizations. Others quote, "In my Father's house are many mansions." The point is that those sects that believe human beings are the pinnacle of the mortal fraction of creation, are going to dislike this a whole lot.

But I think there will be much larger numbers of others who will find this an exhilirating and transforming experience; it's the ultimate de-provincialization of our world view. Think of how much we could learn if there is a decodable message. But even without that, the receipt of a message, clearly of intelligent origin, demonstrates that it's possible to survive our kind of technological adolescence. If we get a message from anybody, since they are clearly far in our technological future, we have learned the hopeful news that it is possible to survive technology.

WHITE: Are we ready for contact? I feel that the more education is done, the more positive it would be, because it would be less shocking and traumatic.

SAGAN: Some people are ready and some aren't. There is no question that some people would find any challenge to their deeply held beliefs shocking, but I believe there are an awful lot of other people who would like to understand what else is possible, and who want their beliefs to be in accord with the universe as it really is.

What are their social institutions? What is their biology like? After billions of years of independent biological evolution, what do they look like? How do they say hello? What about language, music, art, literature? What do they know about science? There are so many exciting questions. I think exhilaration at the prospects of new knowledge would by far dominate our response.

It would be such a spur, such a stimulus to creative endeavor in many fields. It's good to re-think our fundamental assumptions. There is no guarantee that we've gotten everything right. We don't have to slavishly follow every claim made in a radio message from somebody else, but it certainly will stimulate our thinking.

WHITE: And it is potentially a unifying activity.

SAGAN: That is the strongest social value of the SETI search program. Whatever they are like, the differences between them and us is going to be so much larger than the differences between any two human communities on the Earth, that SETI will necessarily have a unifying character. This is true, whether we succeed or we fail. If we succeed, of course, there will be those other guys who are very different from us, and all of us humans will see our brotherhood and sisterhood in a way that's never been recognized before.

And suppose the opposite. Suppose we make a long, sophisticated and unsuccessful search. You can never prove that they are not out there; the most you can show is that they're not easy to find. That would then calibrate something of the rarity and preciousness of life on Earth, and would again serve to unify the human species. It's a no lose program.

WHITE: In *Intelligent Life in the Universe*, you talked about the larger question of the impact of intelligent life on the cosmos. That's another level of impact; if we begin interacting with other intelligent beings, what will that bring in terms of a larger cosmic evolutionary process?

SAGAN: Just think about how one of our ancestors of ten thousand years ago would respond on being brought into the center of one of our large cities, or any high technology enterprise. Most of it would be simply baffling—a noisy, unnatural, incomprehensible dream.

Actually, the rate of technological change for most of those ten thousand years has been slow compared to lately. Now, if you ask, What if we go ten thousand years into the future?, a future in which we haven't destroyed ourselves, then the means and even the objectives of that society might be beyond our ability to understand. And they of course will be human.

What if instead you must figure out some completely *different* kind of life, independently evolved, with different sets of customs, and their technology vastly in advance of our own? What will they do? How could we possibly know? The only thing that seems clear is that we will have a tough time figuring it out.

In *CONTACT*, I proposed a few such activities: One activity for a very advanced civilization might be to criss-cross the Galaxy with a kind of rapid transit system. That was at a much lower level of technology than the ultimate effort I imagined there, which was to seal off regions of the universe to prevent the expansion of the universe into a low-density, star-free, almost empty cosmos. It was a kind of experimental cosmic engineering to reverse the expansion of the universe, at least locally. But needless to say this is the merest speculation.

WHITE: Finally, how would you rate SETI's importance compared to other space exploration activities, in terms of its effect on human beings?

SAGAN: The first thing to bear in mind is that it is phenomenally cheap. That certainly speaks in its favor. Secondly, it is one the few activities where you win whether you succeed or fail. And if you succeed, it's hard to see how there could be any bigger impact on how we view the universe and ourselves. So, if you put it all together, I would say that it is a huge bargain.

Interview with Dr. Jill Tarter,
Project Scientist,
NASA SETI Project
October 24, 1988

WHITE: Let's begin by talking about what is now being done in terms of response to a signal, including work in the protocol area.

TARTER: There is concern in two areas: one is that we want to urge all scientists and researchers involved in this field to be extraordinarily rigorous in their verification procedures because it's likely that any signal detected will be at the limits of sensitivity, and might be quite confusing and difficult to sort out.

We are also concerned that the information might be suppressed because of national competition and perceptions of an advantage from single nations knowing about this. We want to urge that an announcement be made and that people adopt the point of view that any such signal is the property of all mankind, and not just the detecting nation.

We've been working on a protocol, and have a draft which is close to something we can begin taking around to scientific unions. We intend for the document to be signed by the researchers and research organizations that are actually involved. But we would like to have it adopted and formally endorsed by as many scientific unions as possible.

We developed the protocol under the auspices of the International Academy of Astronautics and the International Institute of Space Law. We had two years of joint meetings on this topic, and then last year we had an open drafting session at the meeting in Brighton, England, and about forty people showed up with all kinds of ideas. We used that as the basis for this protocol.

It became clear at this drafting session that there was another issue which most people were eager to discuss, they were very emotional about it, and everyone had their own scenario. That is what you should do with respect to making a reply to a message, and that got into the likely content of a message and so on.

We decided that if we wanted to do anything in a finite amount of time, we couldn't include that aspect in the protocol. So the protocol does not deal with any reply; it specifically says there will be another document dealing with that aspect, which presumably will take some international negotiation.

WHITE: This is interesting in terms of the effect of SETI on society, because you're talking about a new way of interacting at the international level. You're facing an unprecedented question and trying to work out how the planet deals with it. We do have a planetary civilization in many ways, but we don't have any planetary authorities; we're in a "state of nature."

TARTER: Yes. One of the very good papers presented at the joint IAA/IISL meeting in India was by John Logsdon, of the George Washington University Graduate Program in Science, Technology and Public Policy. He considered the policy aspects of SETI, and looked for applicable analogues. The only thing he could come up with were the policies established by Japan and the United States for the announcement of a prediction of a probable earthquake, and an international agreement on the announcement of a nuclear accident. A treaty on the latter was just negotiated and signed. That is particularly germane, because it is action in the absence of a planetary body, but it does have a global effect.

WHITE: How concerned are you about the suppression issue?

TARTER: I don't think it's going to be possible, simply because of the way that scientists work. You're going to want to go to someone independent and get confirmation using a completely different set of equipment. At that point, it becomes impossible to suppress anything, because it's like "telling a secret to one person." As soon as you have to go outside your own organization, that's the beginning of the leakage process. Even if there is an attempt to suppress it subsequent to that, it will probably be impossible to do so.

My main concern is that the leakage potential is there, and it's probably going to impede the scientists trying to do a credible job of verifying the signal. Once the media and the world's attention gets focused on this project, it is going to be an incredible spotlight in which to try and work.

WHITE: Are you sitting down with people like Charles Redmond at NASA Public Affairs to work out some of those issues?

TARTER: We don't have the plan in place to that level of detail yet because our priority has been the scientific steps we are going to require for confirmation. We have a program plan that has been signed off on. We're in the process of developing a project plan, and that will have to include a detailed plan for announcement.

We haven't talked about who will be on the panel of experts. The protocol has tried to deal with that issue, but we're not even sure NASA will sign it as an agency.

We would like to have an international body of experts in place in advance to be called on. But it's terribly difficult to keep such a group intact if they're never used. We have suggested that the SETI Committee within the International Academy of Astronautics act as the keepers of a list of individuals who have said they are willing to serve, and that the list be updated each year.

The hope is that these experts would be some of the first people to get all of the technical detailed information that is available. The purpose would be to get their input on the validity of the information and the interpretation done so far, to help with further analysis and serve as media focal points to get the information out.

WHITE: Part of the question is, "What is the content of the message?"

TARTER: Maybe nothing. But that doesn't matter to me. The fact that you detect a signal answers the fundamental question, "Are we alone?" Even if there is no information content in the signal, or none that can be discerned for a long time, you learn something that is very hopeful with the detection of that first signal.

If it is detected fairly easily, that means signals have to be fairly plentiful in order for a finite project to stumble over them. There can't be just one other civilization out there, because it's unlikely that we will have the right technology and sensitivity required to detect just one other civilization. The only way an early detection can happen is if, on average, civilizations live and continue their technology for a long time, so that you end up with a high probability of having one close by. That's a hopeful indication because it means that you can survive the technical infancy that our planet is experiencing. You don't necessarily blow yourself up once you have the technology to do so.

WHITE: What's your current belief about the presence of fairly abundant intelligent life?

TARTER: I think it's fairly abundant, but that gives me a lot of latitude. I don't place any odds on whether or not the particular project we are trying to start will succeed. I really do appreciate how large a job the search might be, but I fundamentally believe it will succeed at some time.

Starting such a search is another milestone in the maturation of a civilization. In the past, we've undertaken projects that last for many generations—the building of the great cathedrals in Europe or the pyramids in Egypt. But we've only done that sort of thing when there's been a very stong central authority to tell us we have to do it. This particular search represents being willing to start something when we might not see the end, simply because we're curious and we think it's very important to do so.

WHITE: Is it fair to say that the search is valid in itself?

TARTER: Yes. I would be sad not to have succeeded in my lifetime, but if you read our project plan, the goal is not to succeed in detecting a signal. The goal is to succeed in making a systematic search of a precisely defined volume of parameter space, because a goal cannot be something that you cannot claim to attain.

WHITE: I think it's worth doing, in terms of its impact on consciousness.

TARTER: Yes. I do a lot of public speaking and one of the reasons I am always pleased to talk to an audience is that if I can get their attention for an hour, I can try to convince them to have a different perspective, to take a more global look at things, to see themselves slightly differently, and that perspective makes the political boundaries we scratch on the surface of this planet seem less significant. I think that's helpful.

Appendix B

SETI Protocols

Author's Note: The following document is the International Protocol.

Declaration of Principles
Concerning Activities
Following the Detection of
Extraterrestrial Intelligence

We, the institutions and individuals participating in the search for extraterrestrial intelligence,

Recognizing that the search for extraterrestrial intelligence is an integral part of space exploration and is being undertaken for peaceful purposes and for the common interest of all mankind,

Inspired by the profound significance for mankind of detecting evidence of extraterrestrial intelligence, even though the probability of detection may be low,

Recalling the Treaty on Principles Governing the Activities of States in the Exploration and Use of Outer Space, Including the Moon and Other Celestial Bodies, which commits States Parties to that Treaty "to inform the Secretary General of the United Nations as well as the public and the international scientific community, to the greatest extent feasible and practicable, of the nature, conduct, locations and results" of their space exploration activities (Article XI),

Recognizing that any initial detection may be incomplete or ambiguous and thus require careful examination as well as confirmation, and that it is essential to maintain the highest standards of scientific responsibility and credibility,

Agree to observe the following principles for disseminating information about the detection of extraterrestrial intelligence:

1. Any individual, public or private research institution, or gov-

ernmental agency that believes it has detected a signal from or other evidence of extraterrestrial intelligence (the discoverer) should seek to verify that the most plausible explanation for the evidence is the existence of extraterrestrial intelligence rather than some other natural phenomenon or anthropogenic phenomenon before making any public announcement. If the evidence cannot be confirmed as indicating the existence of extraterrestrial intelligence, the discoverer may disseminate the information as appropriate to the discovery of any unknown phenomenon.

2. Prior to making a public announcement that evidence of extraterrestrial intelligence has been detected, the discoverer should promptly inform all other observers or research organizations that are parties to this declaration, so that those other parties may seek to confirm the discovery by independent observations at other sites and so that a network can be established to enable continuous monitoring of the signal or phenomenon. Parties to this declaration should not make any public announcement of this information until it is determined whether this information is or is not credible evidence of the existence of extraterrestrial intelligence. The discoverer should inform his/her or its relevant national authorities.

3. After concluding that the discovery appears to be credible evidence of extraterrestrial intelligence, and after informing other parties to this declaration, the discoverer should inform observers throughout the world through the Central Bureau for Astronomical Telegrams of the International Astronomical Union, and should inform the Secretary General of the United Nations in accordance with Article XI of the Treaty on Principles Governing the Activities of States in the Exploration and Use of Outer Space, Including the Moon and Other Bodies. Because of their demonstrated interest in and expertise concerning the question of the existence of extraterrestrial intelligence, the discoverer should simultaneously inform the following international institutions of the discovery and should provide them with all pertinent data and recorded information concerning the evidence: the International Telecommunication Union, the Committee on Space Research of the International Council of Scientific Unions, the International Astronautical Federation, the International Academy of Astronautics, the International Institute of

Space Law, Commission 51 of the International Astronomical Union and Commission J of the International Radio Science Union.

4. A confirmed detection of extraterrestrial intelligence should be disseminated promptly, openly, and widely through scientific channels and public media, observing the procedures in this declaration. The discoverer should have the privilege of making the first public announcement.

5. All data necessary for confirmation of detection should be made available to the international scientific community through publications, meetings, conferences, and other appropriate means.

6. The discovery should be confirmed and monitored and any data bearing on the evidence of extraterrestrial intelligence should be recorded and stored permanently to the greatest extent feasible and practicable, in a form that will make it available for further analysis and interpretation. These recordings should be made available to the international institutions listed above and to members of the scientific community for further objective analysis and interpretation.

7. If the evidence of detection is in the form of electromagnetic signals, the parties to this declaration should seek international agreement to protect the appropriate frequencies by exercising the extraordinary procedures established within the World Administrative Radio Council of the International Telecommunication Union.

8. No response to a signal or other evidence of extraterrestrial intelligence should be sent until appropriate international consultations have taken place. The procedures for such consultations will be the subject of a separate agreement, declaration or arrangement.

9. The SETI Committee of the International Academy of Astronautics, in coordination with Commission 51 of the International Astronomical Union, will conduct a continuing review of procedures for the detection of extraterrestrial intelligence and the subsequent handling of the data. Should credible evidence of extraterrestrial intelligence be discovered, an international committee of scientists and other experts should be established to serve as a focal point for continuing analysis of all observational evidence collected in the aftermath of the discovery, and also to provide advice on the release of information to the public. This committee should be constituted from representatives of each of the international institu-

tions listed above and such other members as the committee may deem necessary. To facilitate the convocation of such a committee at some unknown time in the future, the SETI Committee of the International Academy of Astronautics should initiate and maintain a current list of willing representatives from each of the international institutions listed above, as well as other individuals with relevant skills, and should make that list continuously available through the Secretariat of the International Academy of Astronautics. The International Academy of Astronautics will act as the Depository for this declaration and will annually provide a current list of parties to all the parties to this declaration.

Contact Verification Committee Description

Author's Note: The Contact Verification Committee is not an official part of the International Protocol or the NASA Protocol, but it does represent one approach to the expert verification committee concept advanced in both protocols. The following description was included in the survey undertaken by Dr. Don Tarter, one of the primary developers of the idea:

The Contact Verification Committee, prior to any alleged discovery of indications of extraterrestrial intelligence, would serve as a source of information to the public, to the media, and to the science community about the nature and progress of SETI searches.

The Contact Verification Committee would have no jurisdiction over the granting of monies or the conduct of any aspect of SETI science. The sole functions of the Contact Verification Committee would be to provide information about SETI searches and serve as an official body for confirmation of the discovery of extraterrestrial intelligence. Individual scientists would also have complete freedom to make any announcement about their research or alleged discoveries.

To properly confirm an alleged discovery, and to minimize confusion and misinformation, the committee would proceed in a set of phased steps. Upon being notified of an alleged discovery, the Contact Verification Committee, if it deems the notification credible, would declare a three-stage alert. Such an alert would merely imply that unusual signals of unknown origin have been received. Upon repeated observations and detailed confirmation procedures, if such signals still seem to be a possible or even probable indication of extraterrestrial intelligence, a stage-two alert would be announced. During this stage, final verification procedures would be used and preparation for massive media inquiries would be made. Upon the final determination by the Contact Verification Committee that there are no explanations for the anomalous signals other than extraterrestrial intelligence, a full and formal announcement of the discovery would be made. This would be known as a stage-one alert. The committee would designate the discovering person(s) or agency and proceed to give its confirming endorsement.

After confirmation of the discovery, the committee would serve as a source of information and organization for the massive number of inquiries that would likely ensue.

From "SETI and the Media: Views from Inside and Out," D. E. Tarter, The University of Alabama in Huntsville, presented at 40th Congress of the International Astronautical Federation, October 7–12, 1989, Malaga, Spain

Author's Note: The following document is an early draft of the NASA post-detection protocol document.

August 17, 1987
National Aeronautics and Space Administration (NASA)
Search for Extraterrestrial Intelligence (SETI)

Post Detection Protocols

Definitions:

Search for Extraterrestrial Intelligence (SETI): The name of a Program administered by NASA whose purpose is to conduct a scientifically verifiable search for evidence of extraterrestrial intelligent life in the universe. This program is currently managed by the NASA Office of Space Science and Applications, Life Sciences Division.

SETI Microwave Observing Project: The name of a project, which is a subset of NASA's SETI Program, that is currently under development to search for pulsed and/or continuous wave drifting and nondrifting signals of extraterrestrial intelligent origin in the microwave window of the electromagnetic spectrum between 1–100 GHz. The SETI Microwave Observing Project uses a bimodal search strategy: a high resolution examination of solar-type stars within 100 light years of earth (the targeted search) and a lower resolution survey of the entire celestial sphere (the Sky Survey).

Detection: As it is used in this document, "detection" refers to the acquisition of an electromagnetic signal which cannot be clearly identified by SETI automated signal verification systems or by on-site operational personnel as either radiofrequency interference (RFI) from a human source, or as an astronomical occurrence, or as another type of extraterrestrial signal. Note that the word "detection", refers only to detection of an anomalous signal. It does not mean detection of a signal of extraterrestrial intelligent origin. Some antic-ipated common "false" detection occurrences include but are not limited to detection of a previously uncatalogued RF source, detec-tion of a previously uncataloged astronomical source, a deliberate hoax, an equipment malfunction.

Post-Detection Protocols: In this document, a description of the United States government policies related to the verification that an electromagnetic signal of extraterrestrial intelligent origin, the dis-semination of information about the signal, and the development of future plans related to determining the message content (if any) of the signal and/or interactions with extraterrestrial civilizations; the series of events, techniques, organizations and personnel needed to verify the signal source, disseminate information related to the signal identification, and plan future activities related to determin-ing the message content (if any) and/or interactions with extraterres-trial civilizations. This does not preclude the development of an internationally recognized post-detection protocol.

Background

As of this writing there is no unambiguous scientific evidence for the existence of extraterrestrial life. However, modern scientific theories imply that extraterrestrial life and extraterrestrial intelligent life are possible. Eight countries have sponsored scientific searches for extraterrestrial intelligent life since 1960: the United States, the Union of Soviet Socialist Republics, Argentina, the Netherlands, France, Australia, Germany, and Canada. Although anomalous signals have been detected, the anomalies could not be verified to be of extraterrestrial intelligent origin. Searches are currently in progress at different locations throughout the world. There is at this time no policy agreed to by the governments of the world for the dissemination of information about a verifiable ETI signal or post-verification activities. Although NASA sponsored and privately sponsored SETI searches are currently in progress in the United States, there is no United States government policy that specifically addresses post-detection protocols.

However, post-detection protocols have been discussed on international levels over the past twenty years, and an emerging consensus emphasizes the following point: the detection of an extraterrestrial civilization is a discovery with such profound implications that it transcends national boundaries and should be the property of all humankind. This is in keeping with the National Space Act of 1958 and the NASA philosophy of providing the widest possible distribution of results from space-related research for the good of humankind. This is also in accord with the 1968 Treaty on the Peaceful Uses of Outer Space to which the United States is a signatory. In the absence of a specific policy dealing with successful verification of an ETI signal, SETI researchers in the United States (and abroad) have adopted this philosophy as an unofficial code of ethics.

It should be emphasized that all current SETI searches are extremely limited. However, when the NASA SETI Microwave Observing Project starts, it will exceed all previous searches combined in the first half-hour of operation. The SETI Microwave Observing Project will be 10 billion times more comprehensive than the sum of all previous searches. Therefore, it is timely to give serious consideration to the development of an Agency policy related to post-detection protocols.

SETI Microwave Observing Project
Draft Post Detection Protocol

The procedures outlined below are based on the principles that all announcements should be prompt and accurate and that a true ETI discovery and any information gained thereby should be disseminated widely and promptly. This is standard policy for the agency.

A signal pattern, detected by the SETI signal processing equipment, can have a variety of causes. It can be:

(a) Radio frequency interference
(b) Equipment malfunction
(c) A distant spacecraft
(d) A hoax
(e) An astronomical source
(f) A true ETI signal

To avoid "crying wolf," as many of these alternatives as possible will be eliminated before any announcement is made.

Alternatives (a) and (b) will be eliminated by a set of automatic verification procedures designed into the system software. These include consulting a roster of known RFI signals, determining that the signal cannot be received on an omni-directional antenna, that the signal shifts frequency as the local oscillator is shifted and is therefore at radio frequency, that the signal disappears when the antenna is pointed off target, that the signal is received at constant strength as the antenna pointing describes a small cone about the true direction, and that the true direction remains fixed on the celestial sphere. An up to date list of known spacecraft directions and emissions will be part of the system data base so the system can automatically test alternative (c).

If no known source is found to be at the coordinates of the signal, the team alerts the SETI Project Office which:

1. **Informs SETI MOP sites of the find to try to ensure continuous reception.**

2. **Informs certain observatories of the find to see if they can verify the signal presence.**

After the source has set, the discovery team:

3. **Replaces the system software by recording the software used for detection and downloading a fresh, protected version for use in the next observation period.**

If no signal is found by 1, 2, or 3, the software that detected the signal is examined for alterations producing a hoax, and the search is resumed. If 1, 2, or 3 succeed in recapturing the signal, the SETI Project Office:

4. **Informs NASA Headquarters of the detection of an anomalous signal. NASA Headquarters senior management is informed and the Administrator takes appropriate action concerning dissemination of the information to other executive branch and congressional officials. NASA Headquarters prepares an appropriate news release, to be used if needed, emphasizing that ETI origin of the signal has *not* been confirmed.**

5. **The Project convenes a meeting of a group of technical experts—astronomers, radio-astronomers, physicists, electronics engineers, etc.—to examine all the data. The experts will be chosen at the time and selection may depend on the nature of the signal. A NASA Headquarters representative will participate in this meeting and will provide updates to senior Headquarters management as appropriate as the meeting progresses.**

The meeting may have three possible outcomes:

I. **The signal is clearly of astronomical origin.**
II. **Further tests or observations are needed to determine its origin.**
III. **The signal is clearly of extra-terrestrial intelligent origin.**

The findings are immediately reported to Headquarters, to the SETI team everywhere, and to key personnel connected with the discovery site. If the findings are case I or II, an IAU telegram is sent announcing the discovery. NASA Headquarters will prepare the appropriate news release. In Case I, a news release is prepared asserting the astronomical nature of the source; in Case II, the release stresses the likelihood of the source being astronomical.

In Case III:

A. The NASA Administrator is informed immediately of the discovery and takes the appropriate action to ensure that the proper executive branch and congressional officials are notified. An announcement is prepared by NASA Headquarters for wide distribution and news conferences will be planned. At the discretion of the Administrator of NASA, a formal announcement by him or the president or both may be broadcast.

B. The scientific and technical results of the discovery will be published as soon as possible in the open literature.

C. One or more of the SETI sites will continue to record the signal. These and other observatory records will be pooled at the ARC data facility for further analysis and interpretation. It is recommended that analysis and interpretation of the signal be performed by an international team of scientists designated by their governments for participation in this activity. This would ensure that the message content of the signal becomes the property of the world.

D. The entire SETI project will be re-examined to see if it should be changed in any way. (The detection of one ETI signal improves the probability that others will be found). This re-examination will involve Headquarters, the SETI Project Office, the SETI Science Working Group, the SETI investigator's Working Group, and the scientific community at large.

Appendix C

The Contact Impact Model: Additions and Variations

The Contact Impact Model is, in its current form, only a starting point for understanding the impact of contact with extraterrestrial intelligence. There are at least three areas of further inquiry to consider:

1. Components to be added to the model;
2. Alternative methods of presenting the model;
3. Using the model to explore the relationship of various factors.

Additional Components

The principal limitation on the model is that it does not directly deal with message content. There are, in turn, two elements to message content: *Clarity* (a point that Don Tarter emphasizes in his work), and *Meaning*.

Clarity is important because an enormous amount of information might be sent to us (Information volume), but if it is not understandable, impact is reduced. Some observers have suggested that we might spend years trying to decipher the information. If the transmitting civilization is far away, it will be impossible to conduct a dialogue with them, and we will receive no assistance in translating their messages.

Meaning is similar to clarity, but relates to what is said. Even if we understand the information, there's always the possibility that it just isn't that profound. If they tell us a lot, but we already know most of what they know, then the impact will be diminished.

Impact will increase as clarity and meaningfulness increase, so these two factors are placed in the numerator of the level one model, resulting in the following new formula:

Impact = *[Parity Difference]* × *[Information Volume]* × *[Clarity]* × *[Meaning]/[Distance]* × *[Time]*

The problem with adding these factors is that they are not easily quantifiable in the same terms as the other components. In fact, they can only be used if we adopt alternative versions of the model, described later in this section.

We also need to do more work on components of the model relating to human society. For example, Michael Michaud cites *social cohesion* as an important element in how well cultures cope with contact. Larry Kaye has also suggested that the state of human civilization at the moment of contact will be critical to impact.

Rate of social learning is another factor to be considered. It is similar to the "species absorption" period cited by Lynn Harper, and is now being studied at Harvard's Kennedy School of Government by Dr. William Clark, in relation to the environment.

Rate of social learning refers to the fact that human society takes a certain period of time to adapt to major innovations. For example, it has taken about forty to fifty years, since the first explosion of the atomic bomb, to create a stable means of managing nuclear weapons. Similarly, it has been twenty years since the first Earth Day, and we are only now seeing early global efforts to manage the environmental crisis.

As the rate of social learning increases, impact will presumably decrease, as humanity quickly absorbs and manages the reality of contact.

Closely related to social learning is *preparation*. The greater the level of preparation for contact, the less the impact will be, because the social learning will have taken place in advance.

Adding these three components to the denominator results in the model being expanded to read:

Impact = [Parity Difference] × *[Information Volume]* × *[Clarity]* × *[Meaning]/[Distance]* × *[Time]* × *[Social Cohesion]* × *[Social Learning Rate]* × *[Preparation]*

As is the case with clarity and meaning, these new factors cannot be added to the model unless other versions are used.

Finally, there is the issue of positive vs. negative impact. The model is neutral on whether impact will be good or bad, because so many complex value judgements are involved. This issue relates back to the question of intent, which was also shown to be more complex than it first appeared. At the moment, it seems premature to include a positive/negative dimension, but we still must realize that it is the most important issue for most people. We must also realize that the model has an inherent bias, in that it is structured so that less impact is generally seen as "better" than more impact.

Version II

Parity, information volume, distance and time can all be stated as actual numbers, such as ten-light years, one hundred years, 50%, etc. Clarity, meaning, social learning, social cohesion, and preparation are more difficult to quantify.

The obvious approach to the problem is to assign rankings of one to three, representing a range from "low" to "high" (social cohesion, preparation, etc.), to all the values, including those that currently have real numbers connected with them (distance, time, etc.)

The formula for Version II of the model would remain the same, then, but the values would be different. Let's look at the example used in Chapter Nine, where contact is made in 1995 with a civilization 12 light-years away, 2000 years ahead of ours, and sending 10% of available information.

In this case, parity difference might be rated as medium (2), information volume as medium (2), distance as low (1) and time as low (1). Let's assume that the clarity of the message is low (1), meaning is low (1), social cohesion is medium (2), social learning is medium (2), and preparation is low (1).

The resulting equation is, then:

$$I = [Pd \times Vi \times C \times M] \, / \, [D \times T \times S1 \times Sc \times P]$$
$$Impact = [2 \times 1 \times 1 \times 1] \, / \, [1 \times 1 \times 2 \times 2 \times 1]$$
$$Impact = 2 \, / \, 4 = .5$$

With this version of the model, the greatest possible impact would be 81 $(3 \times 3 \times 3 \times 3)/(1 \times 1 \times 1 \times 1)$, and the lowest is a fraction of 1, which can be generated by many different combinations. The virtue of this version of the model is that the final numbers are much more manageable and comparable to one another. The weakness is that it requires judgment to make the rankings.

Finally, the two versions of the model can be combined by using the ranked factors as a multiplier of the factors that are not ranked.

Using a combined approach on our earlier example would produce the following result:

$$I = [Pd \times Vi \times C \times M]/[D \times T \times S1 \times Sc \times P]$$
$$I = [2000 \times .1 \times 1 \times 1] / [12 \times 5 \times 2 \times 2 \times 1]$$
$$I = 2,000 / 240$$
$$I = .833$$

Exploring Relationships

It will take time to determine which version of the model is most effective and useful. However, that is less important than using it now as a learning tool. As with the Drake Equation, its greatest value is in sensitivity analysis of different factors.

Let's go back to Version I, for example, and consider some of the extreme cases, which tend to reveal the most about impact. For example, let's compare Scenario 2-A (Contact within or near the solar system) to Scenario 3-B (Contact with a civilization from a galaxy outside the Local Group).

Assume that, in both cases, the contact takes place in ten years, the parity difference is 10,000 years, and the volume of information is 10%. For Scenario 2-A, let's put the distance at .25 light-years, and for Scenario 3-B, let's put it at 10,000,000 light-years.

The result for Scenario 2-A is:

$$I = Pd \times Vi / D \times T$$
$$I = 10,000 \times .1 / .25 \times 10$$
$$I = 1000/2.5$$
$$I = 400$$

The result for Scenario 3-B would be:

$$I = 1,000,000 / 10,000,000 \times 10$$
$$I = 1,000,000/ 100,000,000$$
$$I = .01$$

The difference is enormous, as we would expect. With very little information transmitted, the fact that the aliens are very close to Earth is the deciding factor.

Additional sophistication can be added to the alternative Level Two formula by specifying the "units of knowlede transmitted." In *Intelligent Life in the Universe* (p. 428), Sagan and Shklovskii have estimated that the total knowledge base of humanity can be quantified as $3 \times 10 \uparrow 13$ bits of information. If we assume that human acquisition of this knowledge has taken about 35,000 years since the advent of Cro-Magnon, then the rate of knowledge acquisition (R) has been some $10 \uparrow 9$ bits of new information per year. While reality is unlikely to be so simple, let us further assume that $10 \uparrow 9$ bits per year is a universal rate and that it is also the rate at which knowledge will be accumulated on Earth in the future.

Returning to examples of Tau Ceti and Deneb, and the formula:

$$I \ (Impact) = (PD - D) \times Vi$$

Our result would now read $4 \times 10 \uparrow 10$ bits of new information per year for the transmission from Deneb (40 units $\times 10 \uparrow 9$), and about $2 \times 10 \uparrow 11$ bits of new information per year for the transmission from Tau Ceti (200 units $\times 10 \uparrow 9$.

Investigating the relationship between the Drake Equation and the Contact Impact Model represents another area of valuable research. The answer for the Drake Equation, N, or number of communicating civilizations, is highly relevant to the answer, I, for the model.

As N increase, it seems likely that D, the distance to the nearest communicating civilization, will decrease, as will T, the time between now and first contact. If more civilizations exist in the galaxy, one of them is likely to be closer to us, and we are likely to hear from them sooner.

As N increases, Pd, parity difference, is also likely to increase, because there will probably be more advanced and/or older civilizations present in the galaxy. Similary, Vi, information volume, is likely to be higher, because a larger number of civilizations will have created more knowledge that can be transmitted.

N affects every factor of Level One of the model, then, in such a way as to increase I, or impact. More work needs to be done, however, on how to show the impact quantitatively.

Summary

There is much more to be done with the Contact Impact Model, and readers are invited to critique it, work with it, and improve upon it. It would be especially interesting to see whether it can be generalized, and applied to the impact of other forms of space exploration, as well as other forms of social change.

Appendix D

Addresses of Organizations and Institutions Supporting SETI and/or Space Development

CONTACT: % Department of Anthropology, Cabrillo College, Aptos, CA 95003.

High-Precision Stellar Velocities Group (search for extra-solar planets): % Department of Physics and Astronomy, University of Victoria, P.O. Box 1700, Victoria, B.C., Canada, V8W 2Y2.

International Space University: 955 Massachusetts Avenue, Cambridge, MA 02139.

IAU Commission 51 (Bioastronomy: Search for Extraterrestrial Life): % Department of Astronomy, Boston University, Boston, MA 02215.

NASA/SETI Project: The Ames SETI Office, Mail Code 229-8, NASA Ames Research Center, Moffett Field, CA 94035. The JPL SETI Office, Code 264-802, Jet Propulsion Laboratory, 4800 Oak Grove Drive, Pasadena, CA 91109.

National Space Society: 922 Pennsylvania Avenue, S.E., Washington, DC 20003.

North American AstroPhysical Observatory (NAAPO) (Consortium of volunteers to maintain and operate "Big Ear"): % Otterbein College, Westerville, OH 43081.

Planetary Society: 65 North Catalina Avenue, Pasadena, CA 91106.

Space Studies Institute: P.O. Box 82, Princeton, NJ 08540.

Students for the Exploration and Development of Space (SEDS): Massachusetts Institute of Technology, W20-445, Cambridge, MA 02139.

SETI Institute: 2035 Landings Drive, Mountain View, CA 94043.

Notes

Chapter One

1. Gerald S. Hawkins, *Mindsteps to the Cosmos* New York: Harper & Row, 1983, p. 305.
2. Steven J. Dick, "The Concept of Extraterrestrial Intelligence—An Emerging Cosmology?" in *The Planetary Report*, Volume IX, Number 2, March/April, 1989, pp. 13–17.
3. Michael A. G. Michaud, "The Extraterrestrial Paradigm: Improving the Prospects for Life in the Universe" in *Interdisciplinary Science Reviews*, Volume 4, No. 3, September, 1979, pp. 177–192.
4. Interview with the author, August 20, 1989.

Chapter Two

1. Interview with the author, October 24, 1988.
2. Rick Cook, an author and computer expert, discussed this matter in detail at the 1989 *Contact* conference in Phoenix, Arizona, and in conversations and correspondence with the author.
3. Interview with the author, August 10, 1989.
4. ———, *The American Heritage Dictionary*, New York: Dell Publishing Company, 1983, p. 616.
5. Constable, George, ed., *Life Search: Voyage Through the Universe* Alexandria, Va: Time-Life Books, 1988, pp. 15–28.
6. It's not possible to be absolutely certain even about this assumption. The "panspermia" theory suggests that living organisms, such as bacteria, may travel through space, spreading life from one planet to another. This hypothesis is not taken very seriously by most scientists today, but it has never been disproven.
7. Stringent efforts have also been made to sterilize our own unmanned spacecraft, such as the *Viking* Mars lander, so that we do not contaminate other celestial bodies.
8. *Voyage Through the Universe: The Far Planets*, p. 86.
9. Ibid, p. 61.
10. Carl Sagan, "The Triumph of Voyager," *Parade* Magazine, November 26, 1989, p. 5.
11. Interview with the author, August 17, 1989.
12. Speciation is discussed in some detail in THE OVERVIEW EFFECT, and also in an excellent paper by Ben Finney and Eric Jones, "From Africa to the Stars: The Evolution of the Exploring Animal," which can be found in

Space Manufacturing 1983, vol. 53, *Advances in the Astronautical Sciences*, Proceedings of the Sixth Princeton/SSI Conference on Space Manufacturing, ed. James D. Burke and April S. Whitt.

13. *American Heritage Dictionary*, p. 362.

14. For more details on the "cyberphile" mentality, see the interview with roboticist Hans Moravec, by Ed Regis, *Omni*, August, 1989, p. 74.

15. My own view is that the definition of intelligence must focus on information processing capacity, and must also be observable without knowing what is going on "inside" the system. Current definitions are far too "homocentric."

16. Hoyle is also the leader of a school of thought that assumes microorganisms may be able to live in outer space, moving from planet to planet, "infecting" them with life. I'm indebted to NASA astrophysicist Dr. Richard Hoover for some of the thinking in this section, especially regarding microorganisms and Fred Hoyle's work.

17. In order to devise a detection experiment that would locate intelligence radically different from ourselves, we have to separate the concept of intelligence from the current view, which does not readily recognize higher levels of organization beyond ourselves. Again, this approach might be successful if intelligence is defined as a particular level of information processing capability.

18. Perhaps we don't fully realize that refusing to assume the existence of extraterrestrial intelligence may cause us to overlook it when it appears. If the presence of intelligence makes for a more elegant explanation of phenomena than natural causes would, it should be considered. As Carl Sagan and I. S. Shklovskii put it, "Can some phenomena be understood only if we invoke the intervention of living organisms in technical civilizations?" (*Intelligent Life in the Universe*, p. 476)

19. Otto Struve, *The Universe*. Cambridge, MA: MIT Press, 1962, p. 158. (Thanks to Dr. George Field for the reference.)

Chapter Three

1. Tony Rothman, "What You See Is What You Beget," *Discover*, volume 8, No. 5, May, 1987, p. 91.

2. I. S. Shklovskii and Carl Sagan, *Intelligent Life in the Universe* New York: Dell Publishing Co., Inc., 1966, p. 357.

3. Michael J. Crowe, *The Extraterrestrial Life Debate 1750–1900: The Idea of a Plurality of Worlds from Kant to Lowell* Cambridge: Cambridge University Press, 1986, p. 3.

4. Interview with the author, February 7, 1989.

5. David L. Chandler, "Journeys to the Planets," *Boston Globe*, October 2, 1989, p. 43.

6. Eugene Mallove, *The Quickening Universe: Cosmic Evolution and Human Destiny* New York: St. Martin's Press, 1987, p. 5.

7. ———, "Is Anyone Out There?" *Life*, July, 1989, p. 50.
8. Ibid, p. 52.
9. Ibid, p. 57.
10. "What You See Is What You Beget," p. 91.
11. The concept of "mediocrity" tends to be associated with large numbers of people or things, because it is out of those numbers that "averages" appear. By attacking the meaning of the large numbers in the cosmos, the Anthropic advocates assault the essence of the mediocrity argument.
12. John D. Barrow and Frank J. Tipler, *The Anthropic Cosmological Principle* Oxford: Oxford University Press, 1988, p. 3.
13. Ibid.
14. Ibid.
15. "What You See Is What You Beget," p. 93.
16. *Mindsteps to the Cosmos*, p. 302.
17. Ibid, pp. 302–303.
18. Based on an interview with the author, November 21, 1988.
19. Frank J. Tipler, "Extraterrestrial Beings Do Not Exist," *Physics Today*, April, 1981.
20. Ibid.
21. Ibid. Robert A. Freitas Jr. and Francisco Valdes have proposed a new approach to detection, based on looking for probes. They call it SETA (The Search for Extraterrestrial Artifacts). (See Bibliography for references)
22. Interview with the author, April 29, 1989.
23. "Is Anyone Out There?" p. 50.
24. Ibid, p. 53.
25. Ibid, p. 57.
26. Interview with the author, April 29, 1989 (Jones); and "Is Anyone Out There?", p. 50 (Verschuur).
27. Interview with the author, August 11, 1989.
28. Interview with the author, March 17, 1989.
29. Interview with the author, June 13, 1989.
30. Interview with the author, April 30, 1989.
31. Interview with the author, October 20, 1988.
32. Interview with the author, October 24, 1988.
33. "Is Anyone Out There?", p. 50.
34. Interview with the author, December 7, 1988.

Chapter Four

1. Mary M. Connors, "The Consequences of Detecting Extraterrestrial Intelligence," Paper for "Telecommunication Policy," (undated) p. 15.
2. Interview with the author, April 30, 1989.
3. Interview with the author, February 17, 1989.
4. "Love and Physics," Chet Raymo, Boston *Globe*, September 11, 1989.
5. Robert Graves, *The Greek Myths*: vol. 1. Baltimore: Penquin Books, 1955,

pp. 8–23. I am also indebted to Dr. Madeline Nold for her original insights on creation myths and other myths of all types.

6. ———, The New King James Version of The Bible, Genesis 1. Nashville: Thomas Nelson Publishers, 1984, p. 3.

7. Ibid.

8. The Greek Myths: 1, p. 27.

9. Ibid.

10. Ibid.

11. John Alden Williams, ed., Islam, New York: George Braziller, 1962, pp. 20–26.

12. The motif of having been deserted by God, or the gods, is so strong in human history that it is very tempting to wonder if it has some basis in historical reality. Certainly, if extraterrestrials did come to visit Earth on an exploratory mission, they would probably eventually leave, as did the early sailing ships that stopped for a time in the Pacific Islands.

13. The Bible, Matthew 5:9, p. 875.

14. Ibid, John 18:36, p. 979.

15. As a member of the space development movement, I believe that there is a lot of truth in this view, and I'm not suggesting that it isn't based on evidence. However, it is useful to understand the underlying structure of the belief and its apparent origins.

16. It is also possible that this view of extraterrestrials is correct, and it may well be one of the perceived obligations of the more advanced cultures of the galaxy to help evolving cultures through times of peril. Again the point is to see the relationship of this viewpoint to earlier beliefs.

17. Interview with the author.

18. David Skarr, The Insider (2701 North First Street, #220, San Jose, CA 95131), Interview with Dr. Jill Tarter, February 16–22, 1989, p. 7.

19. I am indebted to Willis Harman, President of the Institute of Noetic Sciences, who was interviewed for this book, for helping me to see this perspective. (Interview of July 11, 1989)

20. Cleveland Amory, "The Best and Worst of Everything," Parade Magazine, New York: January 5, 1986, p. 4.

21. This story appeared in a tabloid paper some time after the Parade article, and I have been unable to confirm it, so it must be viewed with a degree of skepticism.

22. Intelligent Life in the Universe, p. 460.

23. Ibid, p. 457.

24. Ibid.

25. Ibid.

26. Once again, the described behavior is not that far from actual events that have taken place during contact among terrestrial cultures.

27. The Bible, Genesis 6: 1–4, p. 7.

28. Richard Wilhelm, The I Ching or Book of Changes, Bollingen Series XIX, Princeton: Princeton University Press, 1950, p. 329.

29. In *Lightyears*, one of the books claiming to describe extraterrestrial contact, an extraterrestrial tells the human contactee that the aliens are indeed the ancestors of humans, and of the same stock. However, she says they do not reveal themselves openly because whenever they have done so in the past, humans have tried to turn them into gods.

30. Jacques Vallee, *UFO's in Space: Anatomy of a Phenomenon*, New York: Ballantine Books, 1965, pp. 7–8.

31. Ibid, p. 10.

32. It is striking that breeding between humans and "gods" would be prevalent in so many myths and then reappear as part of the UFO phenomenon. *Intruders*, by Bud Hopkins, is one of the more detailed explorations of this theme in the modern era.

33. *Omni* Magazine interview, January, 1980.

34. *The Extraterrestrial Debate*, p. xiii.

Chapter Five

1. *Intelligent Life in the Universe*, p. 380.

2. Giuseppe Cocconi and Philip Morrison, "Searching for Interstellar Communications," *Nature*, 1959, reprinted in *The Quest for Extraterrestrial Life: A Book of Readings*, Donald Goldsmith, ed., Mill Valley, CA: University Science Books, 1980, p. 103.

3. *Intelligent Life*, p. 381.

4. Ibid, p. 388.

5. Ibid.

6. *The Quickening Universe*, p. 151.

7. *Intelligent Life*, p. 391–392.

8. Earth-based sources continue to be more difficult to filter out than natural sources, and the problem is getting worse as human civilization expands.

9. Even today, SETI is one of the least expensive means of exploring the universe because it does not require lifting any payloads from the Earth's surface against the force of gravity.

10. *Intelligent Life*, p. 393.

11. Carl Sagan, *Contact*, New York: Simon & Schuster, 1985. If extraterrestrials concluded that the Nazis were the dominant group on the planet, our contact with them might begin with a deep misunderstanding.

12. *Intelligent Life*, pp. 394–395.

13. Ibid, pp. 471–473.

14. Ibid.

15. Ibid.

16. Ibid.

17. Ibid.

18. The term "entropy pools" is my own. I discussed it with Professor George Field of the Smithsonian Astrophysical Observatory at Harvard, who thought that it had merit conceptually. He mentioned that current studies

of the cosmic background radiation show that it is not in a state of equilibrium, as had been previously believed. Recent experiments show that there is a disequilibrium present, something like my description of "entropy pools." He added, however, that those studying the phenomenon do not consider it to be caused by the presence of intelligence.

19. Jill Tarter, "Summary of SETI Observing Programs," September, 1988, pp. 4, 9.

20. Ibid, pp. 4, 5.

21. Ibid, pp. 6, 9, 11, 13, 15.

22. Interview with the author, October 31, 1989.

23. *NASA SETI Post-Detection Protocols*, Draft of August 17, 1987, p. 3.

24. Lynn Harper, "Search for Extraterrestrial Intelligence (SETI) Microwave Observing Project Fact Sheet," unpublished NASA/MOP document p. 2.

25. Bob Gibson, "SETI: Putting an Ear to the Universe," *Ad Astra* (Washington: September, 1989), p. 14. Also, based on my interview with Bob Arnold, August 30, 1989.

26. "Field Tests of the SETI Systems," Summary prepared by NASA/Ames and JPL SETI Staffs. Ames SETI Office, Mail Code 229–8, NASA Ames Research Center, Moffett Field, CA 94035.

27. ———, "SETI: The Search for Extraterrestrial Intelligence," National Aeronautics and Space Administration, JPL 400–265, Rev. 1, January, 1986, p. 4.

28. Ibid.

29. Ibid.

30. Lynn Harper, "Search for Extraterrestrial Intelligence (SETI) Microwave Observing Project Fact Sheet," (unpublished document) p. 1.

31. Interview with the author, October 24, 1988.

32. Ibid.

33. Interview with the author, August 17, 1989.

34. Interview with the author, October 24, 1988.

35. Ibid.

36. Author's visit to the site, fall, 1987.

37. Interview with the author, August 17, 1989.

38. Paul Horowitz, "A Status Report on the Planetary Society's SETI Project," *The Planetary Report*, July/August, 1987, p. 10.

39. Interview with the author, October 31, 1989.

40. ———, "SETI: The Search for Extraterrestrial Intelligence Program Plan," National Aeronautics and Space Administration (Prepared jointly by Ames Research Center and Jet Propulsion Laboratory, Revised March 30, 1987), p. 2–5.

41. Ibid, frontispiece and overleaf pages of report.

42. ———, "SETI: The Search for Extraterrestrial Intelligence," National Aeronautics and Space Administration, p. 6.

43. Author's interview with Lynn Harper, November 8, 1988.

44. Interview with the author, August 30, 1989.

45. ———, National Aeronautics and Space Administration (NASA) Search for Extraterrestrial Intelligence (SETI) Post Detection Protocols (Draft), August 17, 1987, p. 2.
46. Interview with the author, June 30, 1989.
47. *Post Detection Protocols*, p. 3.
48. Ibid, p. 4.
49. Ibid.
50. Ibid, p. 5.
51. Ibid, p. 6.
52. Interview with the author, November 8, 1988.
53. Michaud interview with the author, December 15, 1988.
54. *Ad Astra*, p. 22.
55. Interview with the author, October 20, 1988.
56. Interview with the author, December 7, 1988.
57. Ibid.

Chapter Six

1. Paul Horowitz, "Searching for Signals from Extraterrestrial Civilizations," p. 5.
2. Interview with the author, October 24, 1988.
3. Interviews with the author, October 31, 1989 (Horowitz), and September 19, 1989 (Morrison).
4. *The Extraterrestrial Life Debate*, p. 73.
5. George Constable, ed., *Voyage Through the Universe: Life Search*, Alexandria, VA: Time-Life Books, 1988, pp. 87–88.
6. Carl Sagan, *Cosmos*, New York: Ballantine Books, 1980, p. 85–110.
7. Robert M. Powers, *Mars: Our Future on the Red Planet*, Boston: Houghton Mifflin Company, 1986, pp. 29–31.
8. *Cosmos*, p. 90.
9. Interview with the author, October 20, 1988.
10. Interview with the author, March, 17, 1989.
11. *Intelligent Life in the Universe*, pp. 409–410.
12. Spacefaring civilizations of the Type II and Type III variety would change this factor. As Rick Cook points out, any star becomes an energy source for such a civilization, not just those that are sun-like. Also, the one-to-one correspondence between stars, planets, and civilizations breaks down.
13. Interview with the author, October 9, 1988.
14. Presentation at Hill and Knowlton Advanced Technology Division, Waltham, MA, and interview with the author, February 7, 1989. The author has been working with Dr. Campbell and with Hill and Knowlton, Inc. to secure private funds for the support of his work. We hope the effort can eventually be extended to other SETI projects.
15. Ibid.

16. Betty Nolley, "Looking for a Wobble," *Ad Astra*, September, 1989, pp. 10–11.

17. An excellent article on the "zone of habitability" by James F. Kasting, Owen B. Toon, and James B. Pollack appeared in the February, 1988 issue of *Scientific American*, pp. 90–97. It is primarily a comparative analysis of Earth, Venus and Mars.

18. Interview with the author, September 13, 1989.

19. Ibid.

20. Dale A. Russell, "Speculations on the Evolution of Intelligence in Multicellular Organisms," in *Life in the Universe*, John Billingham, ed, Cambridge, MA: The MIT Press, pp. 259–270.

21. Ibid.

22. *Cosmos*, p. 249.

23. For, example, if the galaxy is ten billion years old, and the typical lifetime of a technical civilization is one billion years, then about ten percent of all those civilizations that have arisen would still be in existence today (one billion divided by ten billion). With four hundred billion stars in the galaxy, that would be about forty billion (four hundred billion times ten percent) civilizations, if the total were not modified by the other factors in the equation.

In the alternative version of the equation, the lifetime of technical civilizations alone can be used, because it is multiplied by R_x, the mean rate of star formation averaged over the lifetime of the galaxy. If the mean rate of star formation is forty per year, and the average age of a communicating civilization is one billion years, then there would again be forty billion (forty times one billion) civilizations, if their number were not modified by the other factors in the equation.

In both cases, the potential number of civilizations is modified by the other factors in the equation (F_p, N_e, F_l, F_l, F_c), which reduces the final total.

24. *Intelligent Life in the Universe*, p. 418.

25. George Constable, ed., *Voyage Through the Universe: Stars* Alexandria, VA: Time-Life Books, 1988, p. 16.

Chapter Seven

1. Interview with the author, November 8, 1988.

2. *The Quickening Universe*, p. 142.

3. Interview with the author, June 13, 1989.

4. Unpublished NASA/MOP document.

5. These are conversations with a terrestrial past. If Earth were to receive information from an advanced civilization many light-years away, it would be a bit like having a conversation with Earth's future. The implications for social evolution are immense.

6. Interview with the author, June 13, 1989.

7. Based on Funaro interview with the author, September 10, 1989.

8. Poul Anderson, "Contact—1985—Ophelia," *Bateson Project: Cultures of the Imagination Contact III & IV 1985, 1987/Proceedings*, Jim Funaro, Grant McDaniels, Reed Riner, ed., Cultures of the Imagination, Capitola, CA: 1989, pp. 1–10.
9. Ibid.
10. Based on discussions with Greg Barr of the *Contact* organization.

Chapter Eight

1. These works, originally known as the "Foundation Trilogy," are now the "Classic Foundation Series." A different saga, on robots, is evolving so that the two series sometimes appear to interact.
2. Isaac Asimov, *Foundation and Earth*, New York: Ballantine Books, 1986.
3. In an interview with the author, Asimov explained that John Campbell, perhaps the most important science fiction editor at that time, mandated that humans should always win out over extraterrestrials in any conflicts or competitions they might have. Asimov did not want to cooperate with this dictum, so he created two series that had no extraterrestrials in them.
4. Steven J. Dick, "The Concept of Extraterrestrial Intelligence—An Emerging Cosmology?" *The Planetary Report*, Volume IX, Number 2, March/April, 1989, p. 13.
5. Ibid., pp. 15–16.
6. Ibid.
7. Ibid.
8. Because the universe is so vast, we will not be able to prove the absence of extraterrestrial life and intelligence for such a long time that it is almost inconceivable. The Assumption of Mediocrity is easily verifiable in principle, while the Anthropic Principle is far more difficult.
9. Frank White, *The Overview Effect: Space Exploration and Human Evolution*, Boston: Houghton Mifflin, 1987, pp. 4–5.
10. Ben Finney comments on this point in some detail in the interview with him of August 20, 1989, but suggests that the interest in space development and SETI both spring from the same exploratory dimension of human nature.
11. If the evolutionary strategy of the cosmos is to become increasingly more intelligent and aware, the question is whether one or many seed-points is the optimum strategy. Conflicts between many expanding spacefaring civilizations might actually slow the process, while cooperation would speed it up.
12. *Mindsteps to the Cosmos*, p. 2.
13. Ibid, p. 305.
14. Ibid.
15. Ibid.
16. Ibid, p. 298.
17. Ibid, p. 305.

18. Ibid, p. 304.
19. Interview with the author, September 7, 1989.
20. Interview with the author, September 13, 1989.
21. A public debate over an ambiguous signal might reach the wrong conclusions, which is almost as bad as a false alarm.
22. Interview with the author, September 7, 1989.
23. Interview with the author, September 18, 1989.
24. Ibid.
25. Ibid.
26. Lewis Grizzard, "Glasnost Opens Door to UFOs," Springfield *News-Leader*, October 25, 1989, Opinion Page.
27. Ibid.
28. Interview with the author, October 20, 1988.
29. "Listening for Extraterrestrials," a SETI symposium presented by the Planetary Society at Harvard University, September 29, 1985.
30. The twentieth anniversary of *Apollo 11*, in July, 1989, was marked by a number of celebrations, for example, and President Bush used the occasion to reveal his Administration's plan for future space development.
31. Interview with the author, September 18, 1989.
32. D. E. Tarter, "SETI and the Media: Views from Inside and Out," 40th Congress of the International Astronautical Federation, October 7–12, 1989/ Malaga, Spain, p. 2.
33. Ibid, p. 3.
34. Interview with the author, November 20, 1989.
35. D. E. Tarter, "SETI and the Media: Views from Inside and Out," 40th Congress of the International Astronautical Federation, October 7–12, 1989/ Malaga, Spain, p. 4.
36. Interview with the author, May 29, 1989.
37. The problem with the protocol is that agreement on following it must be unanimous for it to work, and there is no unanimity even at this early stage, with relatively few researchers involved.
38. Interview with the author, September 7, 1989.
39. Interview with the author, September 18, 1989.
40. Interview with the author, September 7, 1989.
41. Interview with the author, September 18, 1989.
42. Interview with the author, February 9, 1989.
43. Ibid.
44. Ibid.
45. Ibid. Wilford also suggested that the paper would assign a team of reporters to the story full time for several months, and that their specialties would go beyond those of the science writer alone. There might be a general assignment reporter, religion writer, literary writer, and the entire science department developing different aspects of the story.
46. Ibid.
47. Interview with the author, February 17, 1989.

48. Interview with the author, November 8, 1988.
49. An excellent summary of this theory is available from Creative Initiative/ Beyond War, 222 High Street, Palo Alto, CA 94301.
50. Mary M. Connors, "The Consequences of Detecting Extraterrestrial Intelligence," Paper for "Telecommunication Policy," (undated) p. 3.
51. "Mysterious Sun," editorial in San Francisco *Chronicle*, October 25, 1988, p. A 26.
52. "Consequences," p. 15.
53. Ibid.
54. ——— "Reactions to Man's Landing on the Moon," New York *Times*, July 21, 1969.
55. Ibid.
56. L. Frazer, "A New Age of Enlightenment?", *Ad Astra*, September, 1989, pp. 20–21.
57. John M. Logsdon and Catherine M. Anderson, "Announcing the First Signal: Policy Aspects." Preprint of a paper for 37th Congress of the International Astronautical Federation, Innsbruck, Austria, October 4–11, 1986, Oxford: Pergamon Press, 1986.
58. Interview with the author, June 16, 1989.
59. Ibid.
60. See the Bibliography for details on these books.

Chapter Nine

1. Ben Finney, "The Impact of Contact," Preprint of a paper read at the 37th International Astronautical Congress, Innsbruck, Austria, October 4–11, 1986, p. 1.
2. Ibid.
3. Interview with the author, September 10, 1989.
4. Lance Frazer, "Listening for Life," *Ad Astra*, September, 1989, p. 19.
5. "Is Anyone Out There?," *Life*, Gerrit Verschuur interview, p. 50.
6. Interview with the author, September 10, 1989.
7. Ben Finney, "The Impact of Contact," p. 5.
8. Ibid.
9. It might be argued that the foundation for common ground was laid in the 1960s, when space exploration produced "The Overview Effect," or view of the whole Earth from space.
10. Interview with the author, September 7, 1989.
11. Interview with the author, December 15, 1988.
12. Michael A. G. Michaud, "The Consequences of Contact," *AIAA Student Journal*, Winter, 1977, p. 21.
13. *Contact* conference, Phoenix, Arizona, April 30, 1989.
14. Ibid.
15. Ibid.
16. Interview with the author, November 30, 1988.

17. Interview with the author, February 3, 1989.

18. David Brin, *Startide Rising*, New York: Bantam Books, 1983.

19. Interview with the author, June 13, 1989.

20. Kathleen Best, "Reagan Admits To Curiosity About Alien Attack," St. Louis *Post-Dispatch*, May 5, 1988.

21. For details, see *The Great Chiefs*, Jerry Korn, ed., text by Benjamin Capps, Chicago: Time-Life Books, 1975.

22. I attended a meeting on "Native Science," held near Calgary, Canada, in the summer of 1989, and found the wisdom of the native cultures to be deep and profound. They have learned an enormous amount, through direct experience, about how to survive contact with an "alien culture."

23. Interview with the author, December 15, 1988.

24. Ibid.

25. Interview with the author, November 8, 1988.

26. James E. Lovelock, *GAIA: A New Look at Life on Earth*. Oxford: Oxford University, Press, 1979. In another book, as yet unpublished, I am working on development of "The Cosma Hypothesis," which extends the insight of "The Gaia Hypothesis" to the universe as a whole. The Cosma Hypothesis assumes that the universe is a living information processing system, and that life and intelligence represent different levels of processing capability.

27. Gina Maranto, "Earth's First Visitors to Mars," *Discover* Magazine, May, 1987, Volume 8, Number 5, pp. 28–43.

28. For details on nanotechnology, see Eric Drexler's *Engines of Creation: The Coming Era of Nanotechnology*. New York: Anchor Press/Doubleday, 1987.

29. Carl Sagan discusses the terraforming of Mars in *Cosmos* (p. 110), and Sagan and Shklovskii discuss the terraforming of Venus in *Intelligent Life in the Universe* (p. 467–468). The Dyson Sphere concept, which involves dismantling an entire planet, is a version of terraforming practiced on the scale of a solar system.

30. The classic work on using extraterrestrial materials to build space habitats is Dr. Gerard K. O'Neill's *The High Frontier*. It is also discussed at length in the report of the National Commission on Space, *Pioneering the Space Frontier* (See Bibliography for references).

31. *The Anthropic Cosmological Principle*, p. 659.

32. *Intelligent Life in the Universe*, p. 486.

33. Ibid., p. 476.

34. Ibid., p. 467.

35. Michael A. G. Michaud, "The Final Question: Paradigms for Intelligent Life in the Universe," *Journal of the British Interplanetary Society*, Vol. 35, 1982, p. 131.

36. Ibid.

37. Michael A. G. Michaud, "The Extraterrestrial Paradigm: Improving the Prospects for Life in the Universe," *Interdisciplinary Science Reviews*, Vol. 4, No. 3, 1979, p. 188.

38. Frank White, *The Overview Effect*, p. 93. My idea of the "Universal Overview System" also looks into "The Cosma Hypothesis." It is discussed at length in my next book, *Citizens of the Universe*.

39. For more details on the derivation of these quantities, see *Intelligent Life in the Universe*, p. 130; and *The Global Brain*, p. 57.

40. Interview with the author, November 8, 1988.

Chapter Ten

1. For details on these two approaches, see *The Overview Effect*, Chapter 22 ("Creating the Future") and Michael Michaud's "Sharing the Grand Strategy," *Space World*, August, 1984, pp. 5–9.

2. Interview with the author, November 8, 1988.

3. As David Harper of the NASA SETI project wrote to me in a letter, "The space program has given us a context (pictures of the Earth floating in the black void and the *Voyager* pictures) to think of our place in the solar system, SETI will give us the context to think of our place in the Galaxy."

Bibliography

Asimov, Isaac. *Foundation* (1951), *Foundation and Empire* (1952), *Second Foundation* (1953), *Foundation's Edge* (1982), *Foundation and Earth* (1986), *The Robots of Dawn* (1983), *Robots and Empire* (1985) New York: Ballantine Books.

———— and White, Frank. *Think About Space: Where Have We Been? Where Are We Going?* New York: Walker and Company, 1988.

Baird, John C. *The Inner Limits of Outer Space.* University Press of New England, 1987.

Barrow, John D., and Tipler, Frank J. *The Anthropic Cosmological Principle.* Oxford: Oxford University Press, 1988.

Billingham, John, ed. *Life in the Universe.* Cambridge, MA: The MIT Press, 1981.

Brin, David. *Startide Rising.* New York: Bantam Books, 1983.

Campbell, Joseph, with Bill Moyers, Betty Sue Flowers, ed. *The Power of Myth.* New York: Doubleday, 1988.

Clarke, Arthur C. *2061: Odyssey Three.* New York: Ballantine Books, 1987.

Connors, Mary M., "The Consequences of Detecting Extraterrestrial Intelligence," Paper for "Telecommunication Policy" (undated).

Constable, George, ed., *Voyage Through the Universe* Series including *Life Search, The Cosmos, Stars, Galaxies,* and *The Far Planets.* Alexandria, VA: Time-Life Books, 1988.

Cocconi, Giuseppe, and Morrison, Philip. "Searching for Interstellar Communications," *Nature* 184, 844, 1959.

Crowe, Michael J. *The Extraterrestrial Life Debate: 1750–1900.* "The Idea of a Plurality of Worlds from Kant to Lowell." Cambridge: Cambridge University Press, 1986.

David, Leonard, ed. *Ad Astra: The Magazine of the National Space Society,* September, 1989 (Special Issue on SETI).

Dick, Steven J. "The Concept of Extraterrestrial Intelligence—An Emerging Cosmology?" *The Planetary Report,* March/April, 1989.

Drexler, K. Eric. *Engines of Creation: The Coming Era of Nanotechnology.* New York: Anchor Press/Doubleday, 1987.

Easterbrook, Gregg. "Are We Alone?" *Atlantic Monthly,* August, 1988.

Finney, Ben. "The Impact of Contact." Preprint of a paper read at the 37th International Astronautical Congress, Innsbruck, Austria, October 4–11, 1986.

Freitas, Robert A. Jr., and Valdes, Francisco. "The Search for Extraterrestrial Artifacts (SETA)". Paper presented at 35th Congress of International Astronautical Federation, 8–13 October, 1985. *Acta Astronautica,* Vol. 12, No. 12, pp. 1027–1034, 1985.

Funaro, James J.; McDaniels, Grant, and Riner, Reed, ed. *Contact: Cultures of the Imagination/Proceedings: Contact III 1985; The Bateson Project/ Contact III & IV 1985 & 1987; Contact IV 1987; Contact V 1988;* Capitola, CA, 1989.

Goldsmith, Donald, ed. *The Quest for Extraterrestrial Life.* Mill Valley, CA: University Science Books, 1980. Distributed by Univelt, Inc., San Diego, CA.

Graves, Robert. *The Greek Myths: 1.* Baltimore, Md: Penguin Pelican Books, 1955.

Hawkins, Gerald S. *Mindsteps to the Cosmos.* New York: Harper & Row, 1983.

Horowitz, Paul. "A Status Report on The Planetary Society's SETI Project," *The Planetary Report,* Volume VII, Number 4, July/August, 1987.

Horowitz, Paul; Matthews, Brian S.; Linscott, Ivan; Teague, Calvin C.; Chen, Kok; and Backus, Peter. "Ultranarrowband Searches for Extraterrestrial Intelligence with Dedicated Signal-Processing Hardware," *Icarus 67,* 525–539 (1986).

Hynek, J. Allen. *The UFO Experience: A Scientific Inquiry.* New York: Ballantine Books, 1972.

Lovelock, James E. *Gaia: A New Look at Life on Earth.* Oxford: Oxford University Press, 1979.

Mallove, Eugene T. *The Quickening Universe: Cosmic Evolution and Human Destiny.* New York: St. Martin's Press, 1987.

Marx, George, ed. *Bioastronomy—The Next Steps.* Proceedings of the 99th Colloquium of the International Astronomical Union, Balaton, Hungary, June 22–27, 1987 Dordecht: Kluwer Academic Publishers.

McManus, Jason, ed., "Is Anyone Out There?" *Life Magazine,* July, 1989.

Michaud, Michael A. G. "Interstellar Negotiation," *Foreign Service Journal,* December, 1972; "The Consequences of Contact," *AIAA Student Journal,* Winter, 1977; "The Extraterrestrial Paradigm: Improving the Prospects for Life in the Universe." *Interdisciplinary Science Reviews,* Vol. 4, No. 3, 1979; "Extraterrestrial Politics," *Cosmic Search,* Summer 1979; "The Final Question: Paradigms for Intelligent Life in the Universe," *Journal of the British Interplanetary Society,* Vol. 35, 1982; "Sharing the Grand Strategy," *Space World,* August, 1984; "Detection of ETI—An International Agreement," *Space Policy,* 1989.

Mooney, Richard E. *Gods of Air and Darkness.* Greenwich, Ct.: Fawcett Crest, 1975.

National Aeronautics and Space Administration (NASA). *SETI: The Search for Extraterrestrial Intelligence* (A Program Plan Prepared Jointly by the Ames Research Center and Jet Propulsion Laboratory SETI Program Office), 1987. (Includes an excellent bibliography).

National Commission on Space. *Pioneering the Space Frontier.* New York: Bantam Books, 1986.

O'Neill, Gerard K. *The High Frontier: Human Colonies in Space.* Princeton, New Jersey: Space Studies Institute Press, 1989.

Powers, Robert M. *Mars: Our Future on the Red Planet*. Boston: Houghton Mifflin Company, 1986.

Rothman, Tony. "What You See Is What You Beget." *Discover Magazine*, May, 1987.

Russell, Peter. *The Global Brain: Speculations on the Evolutionary Leap to Planetary Consciousness*. Los Angeles: J. P. Tarcher, Inc., 1983.

Sagan, Carl. *Cosmos*. New York: Ballantine Books, 1980.

Shklovskii, I. S. and Sagan, Carl. *Intelligent Life in the Universe*. New York: Delta Books, Dell Publishing Co., Inc., 1966.

Stapledon, William Olaf. *Star Maker*. Baltimore: Penguin Books, 1937 (reprinted 1973).

Tarter, Don E. "SETI and the Media: Views from Inside and Out." IAA-89-645, 40th Congress of the International Astronautical Federation, October 7–12, 1989, Malaga, Spain (International Astronautical Federation, 3–5 Rue Mario-Nikis, 75015 Paris, France).

Tipler, Frank J. "Extraterrestrial Beings Do Not Exist" *Physics Today*, April, 1981.

Turkle, Sherry. *The Second Self: Computers and the Human Spirit*. New York: Simon & Schuster, Inc., 1984.

White, Frank. *The Overview Effect: Space Exploration and Human Evolution*. Boston: Houghton Mifflin Company, 1987.

Vallee, Jacques. *UFOs in Space: Anatomy of a Phenomenon*. New York: Ballantine Books, 1965.

Vallee, Jacques and Janine. *Challenge to Science: The UFO Enigma*. New York: Ballantine Books, 1966.

Zim, Herbert S.; Baker, Robert H.; and (revised and updated by) Chartrand, Mark R. *Stars*. New York: Golden Press, 1985.

Index